APOMIXIS IN ANGIOSPERMS
Nucellar and Integumentary Embryony

Tamara N. Naumova, Ph.D., Doct.D.
Komarov Botanical Institute
St. Petersburg, Russia

Translated by
I. Mershchikova

CRC Press
Taylor & Francis Group
Boca Raton London New York

CRC Press is an imprint of the
Taylor & Francis Group, an **informa** business

First published 1993 by CRC Press
Taylor & Francis Group
6000 Broken Sound Parkway NW, Suite 300
Boca Raton, FL 33487-2742

Reissued 2018 by CRC Press

© 1993 by CRC Press, Inc.
CRC Press is an imprint of Taylor & Francis Group, an Informa business

No claim to original U.S. Government works

Library of Congress Cataloging-in-Publication Data

Naumova, T. N. (Tamara N.)
    Apomixis in angiosperms : nucellar and integumentary embryony /
T. N. Naumova ; translated by I. Mershchikova.
      p. cm.
    Includes bibliographical references (pp. 71–84) and index.
    ISBN 0-8493-4570-7
    1. Apomixis.   2. Angiosperms--Reproduction.   I. Title.
QK826.N38   1992
582.13'04162--dc20   92-25075

A Library of Congress record exists under LC control number: 92025075

ISBN 13: 978-1-315-89065-4 (hbk)
ISBN 13: 978-1-351-06975-5 (ebk)

Visit the Taylor & Francis Web site at http://www.taylorandfrancis.com and the
CRC Press Web site at http://www.crcpress.com

PREFACE

This publication is devoted to nucellar and integumentary (adventive) embryony, which is a widely spread type of apomixis in nature.

The work is based on original cytoembryological data and critically reviewed literature on more than 250 species from 57 families of angiosperms. Light and electron microscopy techniques have allowed a revelation of important cytological regularities underlying the processes of differentiation in the initial cells of nucellar and integumentary embryos. The processes leading to the formation of asexual embryos and to the development of viable seeds are shown. The occurrence of this phenomenon has been examined within species, families, and among angiosperms in general. A list of species characterized by adventive embryony is compiled, many of them being economically important. An original classification of apomixis and an original approach to differentiation of embryological structures are suggested. They are expected to offer new possibilities for treatment of apomixis.

This book is primarily intended for embryologists but it is hoped that botanists, cytologists, geneticists, and plant breeders will find it helpful. It might be also used in studies of tissue culture. The work presented here was undertaken at the Department of Plant Embryology, the Komarov Botanical Institute, St. Petersburg, and in part at the Department of Plant Cytology and Morphology, Agricultural University, Wageningen, The Netherlands.

I am indebted to many people for their aid in preparing this book. Above all I am indebted to my teacher, Professor M. S. Yakovlev, for support and encouragement.

I am also grateful to my colleagues L.I. Oryol, E.L. Kordyum, M.F. Danilova, A.E. Vassilyev, V.N. Tikhomirov, M.P. Solntzeva, G.Ya. Zhukova, and G.K. Alimova for their most useful comments on the manuscript and moral support they gave me in this undertaking.

In addition I would like to thank my Dutch colleagues, M.T.M. Willemse, H.J. Wilms, and J. Van Went, who were very helpful to me. I wish also to acknowledge technical assistance from L.M. Rotenfeld and N.G. Toutchina.

THE EDITOR

T.N. Naumova, Ph.D., Doct.D., is a senior research scientist at the Komarov Botanical Institute of the Academy of Sciences, St. Petersburg, Russia.

Dr. Naumova graduated from St. Petersburg University in 1965. She obtained her Ph.D. from the Department of Plant Morphology and Embryology, Komarov Botanical Institute of the Academy of Sciences in 1971 and her Doct.D. from the same institute in 1990. After completion of her Ph.D., she was appointed a junior research scientist at the Komarov Botanical Institute in 1971. She became a senior research scientist at the same institute after defending her Doct.D. in 1991 and works there presently.

Dr. Naumova is a member of the Botanical Society of Russia, the International Association of Sexual Plant Reproduction, and the International Network for Apomixis Research (APONET). She was the recipient of an International Agricultural Centre of the Netherlands Fellowship in 1980 and 1991–92.

Dr. Naumova is the author of more than 70 research papers and has presented over 30 lectures at national and international meetings. Her current major research interests include the embryological and cytological investigation of asexual and sexual plant reproduction.

TABLE OF CONTENTS

Chapter 1
General Aspects 1

Chapter 2
Historical Background 3

Chapter 3
**Structural and Functional Aspects of Nucellar and
Integumentary Embryony** 13
I.   Introduction 13
II.   Materials and Methods of Research 13
III.   Microsporogenesis, Microgametogenesis, and Mature
       Pollen Grains 15
IV.   Megasporogenesis, Megagametogenesis, and Mature
       Embryo Sacs 16
V.   Double Fertilization 21
VI.   Endospermogenesis 22
VII.   Initial Cells of Nucellar and Integumentary Embryos —
       Embryocytes 24
VIII. Mitosis of Embryocytes in *Sarcococca humilis* 30
IX.   Development of Sexual and Asexual Embryos 35
X.   Polyembryony 41
XI.   Theoretical Grounds for Nucellar and Integumentary
       Embryony 43
XII.   Adventive Embryos and Embryoids 45

Chapter 4
**Occurrence of Nucellar and Integumentary Embryony and its
Evolutionary Significance** 49
I.   Occurrence of Nucellar and Integumentary Embryony in the
       System of Flowering Plants 49
II.   Nucellar and Integumentary Embryony within Families,
       Genera, and Species of Flowering Plants 51
III.   Evolutionary Significance of Nucellar and Integumentary
       Embryony 56

Chapter 5
**Apomixis and Amphimixis in Seed Production of Flowering
Plants: Classification** 57

Chapter 6
A List of Species with Nucellar and Integumentary Embryony 63

References 71

Abbreviations/Illustrations 85

Index 141

Chapter 1

GENERAL ASPECTS

Reproduction is a fundamental problem of theoretical botany and at the same time is of primary practical importance. Angiosperms are reproduced in most cases by seeds. The embryo and endosperm of a seed can be formed in two ways: by amphimixis or by apomixis. In the case of amphimixis the embryo and endosperm are produced as the result of double fertilization. No fusion of the gametes is observed in apomixis; formation of the embryo, endosperm, and embryo sac is rather diverse, but it also leads to production of viable seeds. Apomictically originated embryos and developing plants have genetically inherited properties different from those of amphimictic ones.

Apomixis occurring in nature contributes to seed productivity and genetic stability of the flowering plants. Embryological data on apomixis are scanty. Terminology of apomixis is complicated and confusing and classifications of this phenomenon are numerous, but none of them is universally accepted. Apomixis is most diverse among angiosperms compared to all other representatives of the plant and animal kingdoms. Underlying causes are a peculiar development of generative structures of the flowering plants: structural and functional features of the ovule, spore type of meiosis, and potentialities of the embryo sac cells in respect to their reproductive functions.

Nucellar and integumentary embryony (collectively called adventive embryony) is one of the most widely spread types of apomixis in nature; the embryo is formed from separate nucellus or integument cells but is developed in the embryo sac and produces a viable seed. Polyembryony, the formation and development of several embryos in a single seed, is typical of adventive embryony. It is generally caused by the formation of several initial cells (embryocytes) which can be developed into nucellar or integumentary embryos.

Adventive embryony has been studied inadequately. Embryological investigations carried out so far are for the most part fragmentary. Information on adventive embryony cited in publications on general embryology and apomixis does not provide a comprehensive picture of this phenomenon.[1-9] The role of adventive embryony in the system of apomixis and angiosperm reproduction remains undefined, due to the limited knowledge available and the unique characteristics of this phenomenon. Some investigators exclude adventive embryony from apomixis and regard it as equivalent to vegetative multiplication[6,10-13] or consider adventive embryony a remarkable means of vegetative propagation,[14] sometimes comparing it to cleavage polyembryony;[15] other researchers included it in apomixis, but often treat it separately from its other types.[16-27] There is no unanimous opinion among scientists on this issue. Data on the evolutionary significance of adventive embryony are practically lacking.

1

Many questions associated with embryology of flowering plants which possess this type of apomixis remain unresolved. Thus, adventive embryony is not clearly understood. Therefore, we believed it necessary to trace carefully all the successive stages of development of species with adventive embryony.

Interest in adventive embryony is not surprising because in the plant kingdom it can occur only among angiosperms and is connected with the realization of reproductive potentiality, not only by specific cells of the ovule (megasporocytes), but also by other nucellus or integument cells (embryocytes). This property illustrates totipotentiality exhibited by the ovule cells which can be widely used in practice. Adventive embryony becomes of current significance because it is found in a number of economically important plants belonging to genera such as *Citrus, Mangifera, Malus, Ribes, Beta,* and some genera of the family Poaceae. For practical purposes, it can be used to maintain heterosis, to stabilize hybrids, and to produce virus-free and disease-resistant forms of plants. Moreover, the species with adventive embryony might be employed as promising objects for works in tissue culture.

The aim of the present work is to develop a comprehensive characteristic of adventive embryony as one of the modes of seed reproduction in angiosperms and to reveal main cytoembryological regularities that underlie this phenomenon. This highlights major problems of embryological development of the species with nucellar and integumentary embryony on the basis of original data and a survey of literature with respect to representatives of more than 250 species from 57 families of angiosperms. Ultrastructural investigations on initial cells of adventive embryos (embryocytes) have been carried out for the first time. A list of species with nucellar and integumentary embryony is compiled which also includes other types of apomixis and references. Occurrence of this phenomenon is examined within species, genera, families, and among angiosperms in general. An original classification of apomixis and amphimixis is suggested.

As a result of the present studies, some new principles are proposed with respect to the development of male and female generative structures, development of sexual and adventive embryos, and peculiarities of endospermogenesis. Ultrastructural investigations offered a means to reveal typical features of embryocyte differentiation and to gain new insight into adventive embryony and the problem of apomixis in general. It was shown that structural transformations of the embryocyte protoplast in many aspects were similar to those of the protoplasts of other embryonic structures developing in the ovule. The research results reveal evidence for common regularities in differentiating embryological structures in the case of both apomixis and amphimixis. Apart from the above practical purposes, these data might also be helpful in solving theoretical aspects of the problem of seed production.

Chapter 2

HISTORICAL BACKGROUND

Investigations of angiosperms dating to the 19th century showed that seeds could be produced without pollination and fertilization. Smith, an English scientist from the Royal Botanical Gardens at Kew, was the first to discover (in 1841) normally formed and viable seeds of a female specimen of the dioecious Australian plant *Alchornea ilicifolia* in the absence of male individuals.[28] This report and information on polyembryony encouraged scientists to undertake wider and more profound studies of generative structures.

The first embryological studies of plants characterized by polyembryony were made by Strasburger.[29] He found that additional embryos could be formed in the seed without fertilization, directly from nucellus cells adjacent to the embryo sac. He called such embryos adventive (Adventivkeime) embryos and the phenomenon as a whole adventive embryony. It was described by Strasburger using the following species: *Funkia ovata, Alchornea (Coelebogyne) ilicifolia, Citrus aurantium, Mangifera indica, Euonymus latifolia, Gymnadenia conopsea* and *Nothoscordum fragrans*. Adventive embryony was the first type of asexual reproduction discovered in flowering plants. In this case separate cells of nucellus were capable of functioning as a zygote. Apart from adventive embryony, other instances of embryo formation in angiosperms without fertilization were later described. For those cases and the phenomenon as a whole, the term apomixis was suggested to denote the formation of a new organism without fusion of the gametes.[30,31] Vegetative multiplication was also included by Winkler in the notion of apomixis. Somewhat later the term agamospermy was introduced that covered all cases of reproduction via seeds without fusion of the gametes. These terms are still widely used and we employ them along with other necessary notions generally accepted in this field. The term adventive embryony is used in the present work, as in works by many other researchers, as an equivalent to nucellar or integumentary embryony. The early works by Strasburger[29] in the field of adventive embryony were followed by a series of further studies. This phenomenon was soon observed in many plants: *Ficus roxburghii, Erythronium americanum, Allium odorum, Opuntia vulgaris, O. ficus-indica, O. rafinesquii, Spiranthes cernua, Alchemilla* sp., *Euphorbia dulcis, Colchicum autumnale, Wikstroemia indica, Mangifera indica, Eugenia jambos, Zanthoxylum bungei, Citrus aurantium, C. nobilis,* and *Smilacina racemosa*.[30,32-45] Most of these early embryological works stated primarily only the presence of adventive embryos in the seed. Observational results had much in common with the pictures described by Strasburger,[29] though some additional data were naturally obtained. Ganong[35] found in particular that the egg cell of *O. vulgaris* degenerated and all embryos of the seed were of nucellar origin. It was shown in the example of some species of *Alchemilla* that nucellar embryos could start developing prior to endosperm and their development could be unrelated to pollination.[39] In the course of the study of *Euphorbia dulcis*, the development of embryos from unfertilized egg cell and synergids was observed along with adventive embryo formation.[40]

During this period embryological investigations of economically important plants characterized by adventive embryony were also started: *Mangifera, Euphorbia,* and *Citrus.*[42,44] The work by Osawa[44] deserves particular notice. He closely examined the development of two sterile varieties of the genus *Citrus:*Washington navel orange (*C. aurantium*) and Unshiu (*C. nobilis*); in parallel some fertile representatives of the genus were under study. Various anomalies were found in the formation of the embryo sacs and pollen grains, though no differences in the development of the ovule and its envelopes were encounted in fertile and sterile forms. Osawa's drawings were executed so carefully and accurately that even nowadays they are presented in manuals on plant embryology.

One of the first Russian works in the field of adventive embryony belongs to Tretjakov[33] from the Moscow Botanical Garden who studied *Allium odorum.* The work was later continued by Modilewsky,[46,47] who showed that embryos of this species could arise from an egg cell, synergids, antipodal or integument cells. It was noted that embryos could develop only in the presence of the endosperm which arises from fertilization. Due to possible formation of embryos from both nucellus and integument cells, adventive embryony was subdivided into nucellar and integumentary embryony.[46-48] These concepts were first introduced into the classification by Webber.[49]

Further studies revealed the occurrence of adventive embryony in many representatives of flowering plants. We have compiled a list of the angiosperm species characterized by nucellar and integumentary embryony on the basis of major relevant publications and our research data.

There are no comprehensive publications on adventive embryony at present. Basic data on this phenomenon are presented in works dealing with apomixis in general terms or in outlines of plant embryology. Any interested reader will see how confusing the problem is but will fail to find definite answers in these publications. Most specific works on embryology of taxa with nucellar and integumentary embryony which we will discuss later have been unfortunately disregarded as a general characteristic of the phenomenon.

The first classification of apomixis was given by H.Winkler[30] Adventive embryony was not discussed. Of significant interest is a later work by this author[14] where he examined separate forms of apomixis, its nature and origin, and its biological significance. Winkler was the first to draw attention to differences in alternations of nuclear phases and generations in the case of amphimixis and apomixis. Since this point was mentioned in later works by other investigators, we shall treat it at some length. Starting from the principles based on alternations of nuclear phases in the plant ontogenesis, Winkler subdivided apomixis into three types: (1) vegetative multiplication, (2) apogamy, and (3) parthenogenesis. He places reproduction by means of spores, off-springs, bulbs, etc. into vegetative multiplication. Winkler regards adventive embryony as a remarkable means of vegetative multiplication since a new sporophyte (adventive embryo) arises from an old sporophyte (the ovule cells) and as a consequence there is no alternation of nuclear phases and generations. Adventive embryony, in Winkler's opinion, is characterized by the formation of "offsprings" (propagation sprosse) within the ovule directly from its cells; these offsprings then penetrate into the embryo sac and are transformed into adventive embryos which are considered as "runners" of the sporophyte. Though taking into account that in adventive embryony, like apogamy and parthenogenesis, seeds are still produced, Winkler applies the collective term "agamospermy" to all three types of apomixis. Apogamy includes the cases of apomictic origin of a sporophyte (embryo) from vegetative cells of the game-tophyte (synergids). Apomictic production of a sporophyte from the embryo cell (egg cell) is referred to by Winkler as parthenogenesis. Two latter types of apomixis are subdivided into reduced and unreduced ones.

The classification of apomixis developed by Winkler is illustrated when a phenomenon is examined using one basic criterion: alternations of generations and nuclear phases in plant ontogenesis in the case of amphimixis and some types of apomixis. However, it can be seen from the above that the author encounted certain difficulties in the treatment of apomixis. They are primarily concerned with adventive embryony because on the one hand it is regarded as a form of vegetative multiplication and on the other hand as a form of agamo-spermy.

In 1918, Ernst[16] published the work where the following three types of reproduction were distinguished: apomixis, amphimixis, and vegetative multi-plication. Apomixis was divided into two types: (1) apomixis with an alterna-tion of generations, including apogamy (= diplospory — development of the unreduced embryo sac from a megasporocyte), apospory(= development — of the unreduced embryo sac from a nucellus cell), apogamety (= parthenogenesis — development of the embryo from egg cell), and somatic apogamy (= apogamety — development of the embryo from synergids) and (2) apomixis without an alternation of generations that integrates nucellar embryony, par-thenocarpy, apoflory, and apocarpy.

Examination of nucellar embryony jointly with the above anomalies which do not produce viable seeds is probably not quite justifiable. However, inclu-sion of nucellar embryony in apomixis deserves notice. Another new point of the classification was subdivision of apogamy and nucellar embryony into induced and autonomous forms that are dependent on or independent of pollination. This conclusion was inferred from very few first works on plant embryology. However, even these limited data were controversial and views on the dependence of nucellar embryony on pollination were a matter of debate.[29,39,44] We believe that subdivision of apomixis into autonomous and induced forms suggested by Ernst was not completely justified. Later works have failed to provide a definite answer. In spite of that, apomixis forms remain divided into induced and autonomous in most subsequent classifications.

One of the first and most comprehensive publications on plant embryology[1,50] gives a very brief description of adventive embryony and its content practically does not exceed that volume of data first reported by Strasburger.[29]

Another work on classification and terminology of apomixis was published by Fagerlind;[18] it concerned different means of reproduction in higher plants. The author considered the formation of a reduced gametophyte arising as the result of meiosis and of an unreduced gametophyte produced in the course of abnormal meiosis or apomeiosis. Means of gametophyte formation during apomeiosis are divided into somatic and generative apospory, semiapospory and diplospory. The work also indicates two possible courses of embryo development from cells of the unreduced gametophyte: (1) as the result of their fertilization and (2) as the result of apogamety (from synergids and antipodals) and parthenogenesis (from the egg cell). A significant advance was made by Fagerlind in the treatment of the formation of the unreduced gametophyte. He noted that this process was responsible for certain types of abnormal meiosis such as semiheterotypic, pseudohomotypic, and apohomotypic. However, to place apospory into the types of apomixis occurring as a result of apomeiosis is not quite accurate.

Nucellar embryony is incorporated by the author into agamospermy as an independent branch of the process of seed reproduction. Other forms of apomixis related to unreduction of gametophyte are incorporated into the concept of agamogony. Vegetative multiplication is regarded as an element of apomixis. Thus, Fagerlind, like other investigators dealing with classification of apomixis, refers nucellar embryony to apomixis as an independent and isolated branch. Species with adventive embryony retain the capability to produce seeds; only this probably prevented the author from qualifying it as vegetative multiplication.

The phenomenon of apomixis in angiosperms was also discussed by Stebbins.[10,51] The author suggests dividing apomixis into agamospermy (reproduction by means of seeds) and vegetative multiplication. He states (page 383) that the simplest method of reproduction is adventive embryony in which embryos develop directly from diploid, sporophytic tissue of the nucellus or ovule integument and the gametophyte stage is completely omitted. All other forms of agamospermy, excluding adventive embryony, are regarded as gametophytic apomixis. Acknowledgment of gametophytic apomixis is widely accepted at present.[12,52,53] In-depth analysis of the aspects of species formation and formation of agamic complexes by apomictic plants is found in Stebbins' work. However, the author applies his findings only to species capable of gametophytic apomixis and consequently he excludes adventive embryony.

One of the most fundamental studies on apomixis was and still is, in our opinion, a monograph by Gustafsson.[20-22] It consists of three parts dealing with the mechanism of apomixis, the causal aspects of apomixis, and biotype and species formation, respectively. The author suggested an original system of classification of apomixis forms. He subdivided apomixis into agamospermy (reproduction by means of seeds) and vegetative multiplication. Three independent lines are distinguished in agamospermy: diplospory, apospory, and adventive embryony. The present classification treats adventive embryony on equal terms with all other forms of apomixis which are included in agamospermy. All these forms of apomixis are subjected to careful examination in which data on cytology, embryology, and relevant issues of general biology are accounted for. Still, adventive embryony is discussed on the example of a few species of *Citrus, Poncirus, Alnus, Sarcococca,* and *Allium.* The author draws the conclusion that the present form of apomixis more readily occurs at the diploid level. This view is shared by investigators at present.[12,53] Gustafsson does not make any analogies between adventive embryony and vegetative multiplication. The classification of apomixis suggested by Gustafsson is logical, illustrative, and the terminology he used is simple and accepted at present.

The classification of reproduction modes in angiosperm was developed by Modilewisky.[4] He followed Winkler[14] in regarding adventive embryony as a form of vegetative apomixis; in the case of adventive embryony, reproduction occurs, in Modilewisky's opinion, without involving the gametophyte and is characterized by isolation of some parts of the mother plant, as takes place in vegetative multiplication. Later, the author changed his view on adventive embryony; he revised the statements on the role of a gametophyte and identity of adventive embryony to vegetative multiplication.[5] One of the advantages is binar terminology, suggested by the author, which takes into account how the embryo sac and embryos are formed.

A characteristic viewpoint on apomixis was held by Khokhlov,[23] who gave much consideration to studies of this phenomenon. He thought that a gradual reduction of the gametophyte occurred in apomictic plants when one or several morphobiological phases (author's terminology) such as spore, archesporium, zygote, etc. are omitted from the typical life cycle. Nucellar and integumentary embryony was referred to as apogametophytic sporophyty. He regarded it as one of the most advanced forms of apomixis in which both the gametophyte and spore are reduced. On this basis an original classification and terminology were developed. Khokhlov assumed that apomixis was at a high level in the evolution of flowering plants.

This view was sharply criticized by a number of investigators.[5,54] Levina indicated in her work[54] that no reduction of the gametophyte and sexual process can be traced in the evolution of a wide variety of plants, both lower and higher. Khokhlov published a number reviews devoted to evolutionary and genetic aspects of apomixis. In addition, he compiled comprehensive lists of apomictic species. Voluminous scientific literature gathered by Khokhlov has undoubtedly contributed to increasing interest in apomixis research.[55-57]

Maheshwary[2] gave a reasonably detailed account of adventive embryony, practically the first description of its kind. Adventive embryony is a type of apomixis. His analysis of a considerable number of works on various genera and species of angiosperms (*Eugenia jambos, Eugenia* sp. *Capparis frondosa, Mangifera indica, Hiptage madablota, Nigritella nigra, Zeuxine sulcata, Alchornea ilicifolia, Zygopetalum maskayi, Opuntia aurantiaca, Sarcococca ilicifolia, S. pruniformis,* and numerous representatives of *Citrus* and *Euphorbia*) has revealed major features of this phenomenon:

1. The cells that participate in the formation of adventive embryos become cytoplasmically dense.
2. A separate cell or a small group of cells can give rise to the embryo.
3. Zygotic and adventive embryos can concurrently develop in the seed; in this case adventive embryos are located at the lateral side of the embryo sac and are devoid of a suspensor.
4. Adventive embryos are developed, provided the endosperm is present; the only exception is *O. aurantiaca*.[58]
5. Morphological features of adventive and zygotic embryos are very similar.
6. Nucellar seedlings have several morphological features typical of zygotic seedlings (the presence of needles in *Citrus*, etc.).

Nygren[3,59] published two works on apomixis in angiosperms where Gustafsson's terminology was accepted. Using the research results of other workers and in part his own findings on the chromosome numbers and embryology, Nygren analyzed the occurrence of sexual and apomictic forms in angiosperms with particular emphasis on the families Asteraceae and Poaceae. There is a brief examination of adventive embryony involving the limited data on *Citrus*. Of interest are the data on endosperm development in agamospermous species concerning chromosome numbers as well as the effect of light regime and other factors on the frequency of aposporous embryo sac formation.

Of particular notice is the publication edited by Maheshwari,[60] two chapters of which are devoted to apomixis[11] and polyembryony,[61] respectively. Adventive embryony, which Battaglia called "sporophytic multiplication", is excluded from the apomixis category and is regarded as equivalent to vegetative multiplication. Battaglia states (page 221 in Reference 11) that apomixis, in contrast to amphimixis, signifies the production of the sporophyte from the gametophyte without sexual fusion. He follows Winkler[14] in placing much emphasis on alternation of developmental phases in plant ontogenesis (sporophyte-gametophyte) as can be seen from Battaglia's classification of apomixis. Apomixis is subdivided into two large groups: (1) heterophasic reproduction and (2) homophasic reproduction. The first type of reproduction is related to an alternation of developmental phases while the second is not connected to the above process and is called "multiplication". According to Battaglia, heterophasic reproduction includes all cases of apomixis, while homophasic reproduction comprises all cases of vegetative multiplication and adventive embryony (seed reproduction). We hold the view that such an integration would require solid grounds. However, the problem of structural and functional transformations during an alternation of developmental phases in ontogenesis has not been virtually discussed in scientific literature. There are other aspects of this problem. It is known in particular that embryos arising as the result of diploid parthenogenesis and apogamety have diploid, maternal chromosome sets and at the same time adventive embryos formed from nucellus or integument cells also have diploid, maternal chromosome sets. Moreover, all these embryos, whatever their origin, are developed in the embryo sac with subsequent production of seeds. A question then arises if these differences in the nature of ontogenesis are really so important because in either case the results are similar. Our concept of this point has been published recently[62,63] and will be discussed later. Of particular note in Battaglia's classification is the separate treatment of methods of embryo sac and embryo formation with the emphasis on the nature of abnormal meioisis, as it was done by Fagerlind.[18] It should also be mentioned that Battaglia described for the first time the phenomenon of semigamy, which is reflected in his classification of apomixis. Battaglia's observations gave a new impetus to apomixis research but eventually led to an even greater diversity of opinions.

Adventive embryony is considered in Reference 61. The authors modified the classification of polyembryony suggested by Lebegue;[64] they distinguished two groups: A — true polyembryony and B — false polyembryony. Along with true polyembryony, connected with cleavage of the zygote, was additional synergid and antipode embryo development. Group A comprises nucellar and integumentary embryony, exemplified only by well-known cases (*Citrus, Mangifera*). We assume that such a limited treatment of adventive embryony does not allow investigation of its nature because the main regularities of this phenomenon are substituted by their consequences, i.e., formation of several embryos that might be a specific case of adventive embryony.

From 1950 to 1980, a series of works on plant embryology appeared in which adventive embryony was discussed. An original classification of polyembryony was suggested by M.Yakovlev.[65,66] It takes into account the origin and life histories of generative structures of higher plants. Polyembryony is divisible into gametophytic and sporophytic. Nucellar and integumentary embryony are referred to as the sporophytic group together with zygotic and proembryonal polyembryony (cleavage polyembryony).

A valuable work in the field of embryology and apomixis was performed by Poddubnaja-Arnoldi.[7,17,67] Her classification[7] gives five types of apomixis, including nucellar and integumentary embryony (induced and autonomous).

The arrangement of apomixis types in the classification demonstrates most effectively their difference from a normal sexual process and can provide useful information on evolutionary stages of apomictic forms. The latter issue is controversial and the existence of evolution in apomictic forms is not universally recognized. There is a brief examination of adventive embryony with reference to another work[68] where this phenomenon is treated at some length. Poddubnaja-Arnoldi proposes to reject the term "adventive embryony" as unfitting. She indicates that all embryos arising from the embryo sac cells (i.e., embryos from the egg cell, synergids, and antipodes) as well as nucellar and integumentary embryos might be called adventive. A similar interpretation of this term can be found in some reference books.[69] We believe it more reasonable to apply this term only to the embryos which are formed from nucellus or integument cells, as was originally suggested by Strasburger[29] and then became a popular notion.

A significant contribution to apomixis research was made by Petrov.[24,25] He carefully analyzed theoretical principles and provided evidence to support the hypothesis of genetically controlled apomixis. He also examined the main forms of apomixis and their evolutionary value. Petrov developed a system of classification; separate chapters were devoted to experimental research of apomixis with respect to plant breeding.

Apomixis in Petrov's classification is divided into agamospermy and vegetative multiplication. Agamospermy is subdivided into regular and irregular apomixis. The latter comprises haploid parthenogenesis and apogamety. Regular apomixis incorporates diplospory, apospory, and adventive embryony; they can occur autonomously or pseudogamously (i.e., in the presence of endosperms produced by fertilization or without fertilization). Petrov examined the well-known works on adventive embryony but, like many investigators, he did not analyze the course of embryological processes.

The most recent monograph by Petrov[9] should be mentioned. He distinguished various groups of apomixis elements that, in his opinion, control one or more forms of apomixis. Much consideration is given to irregular forms of apomixis due to their economic significance. They are based on the embryo development from haploid egg cells, synergids, and antipodes in the natural conditions, as well as from microspore embryoides in culture. Petrov believes that the characteristic feature of adventive embryony is the embryo formation from offshoots of the somatic tissue cells which penetrate into the embryo sac and give rise to adventive embryos (page 47 in Reference 9). This concept of adventive embryony does not practically differ from that of Winkler.[14] Petrov also assumes that normal development of apomictic embryos in unfertilized embryo sacs is possible in the absence of an endosperm. To confirm this statement he only cites Archibald's work,[58] whose findings seemed doubtful to a number of investigators.[61] At the same time the development of adventive embryos and the role of endosperm were observed in a number of works.[70-78]

A monograph on apomixis was written by the well-known German investigator Rutishauser.[6] This work treats in particular the anomalies of meiosis observed in apomictic species of flowering plants. It also discusses in more detail relevant problems of adventive embryony. An attempt is made to summarize information available on the processes of pollination, gametogenesis, and endosperm development in species possessing this type of apomixis. He found to his regret that data on the origin and development of generative structures and on their ploidy and dependence between emergence of initial cells and pollination were practically lacking. The question arises whether the presence of an embryo sac is essential for the development of a nucellar embryo or if it is a remnant feature inherited from its ancestors. Rutishauser makes the point that adventive embryony can be understood if more comprehensive and consistent data would emerge.

Following Battaglia,[11] a classification of apomixis was developed by Solntzeva;[26] some years later it was revised and enlarged.[27] Along with embryo sacs and embryos, the classifications specify possible disruptions of meiosis which lead to nonreduction of the chromosome numbers of the embryo sac cells. This indicates possible sources of genetic nonhomogeneity of developing embryo sacs and embryos in the case of aneuspory (diplospory), which in natural conditions gives rise to agamic complexes by apomictic plants. The author suggests using the term "hemigamy" instead of "semigamy" to be more consistent linguistically. Solntzeva, in contrast to Battaglia, included adventive embryony in apomixis, owing to the fact that development of adventive embryos is closely related to the embryo sac (gametophyte). She probably considered this factor to be more important than the absence of alternations of generations in the case of adventive embryony. As indicated above, a number of research workers thought the latter observation to be a major obstacle to inclusion of adventive embryony in apomixis.

In recent years investigators have been more inclined to support Stebbins[51] in subdividing apomixis into gametophytic apomixis and adventive embryony.[12,52,53] They include diplospory, apospory, parthenogenesis, apogamety, and hemigamy in the category of gametophytic apomixis. They treat adventive embryony as an independent case of agamospermy, which is not included in gametophytic apomixis. Thus, Grant[12] subdivides modes of seed reproduction in flowering plants into A — a normal cycle of sexual reproduction, B — gametophytic apomixis, and C — adventive embryony. In apomixis research some workers take into account endosperm production because endosperm can be formed either as the result of fertilization or due to the fusion of polar nuclei without fertilization.[53] Gametophytic apomixis will be discussed later.

In the monograph devoted to embryology of angiosperms,[79] adventive embryony is discussed in the chapter on polyembryony.[8] Among forms of polyembryony presented in the classification, the authors discern sporophytic (adventive) polyembryony, which is subdivided into nucellar and integumentary types. There are brief descriptions of each form. It is noted that nucellar polyembryony can exist either in the presence or the absence of the gametophyte; the latter occurs in *Cortaderia jabata*. Data on the induced and autonomous development of nucellar embryos of *Citrus* are discussed. The authors draw the conclusion that favorable development of nucellar embryos is dependent on egg-cell fertilization and endosperm formation, whether it is sexual or asexual; in rare instances it can take place even if fertilization of polar nuclei and egg cell has failed. There is a brief analysis of works concerning embryological development of some representatives of the families Rutaceae, Anacardiaceae, Myrtaceae, Cactaceae, and Orchidaceae. The authors assume that integumentary embryony is always achieved, irrespective of egg-cell fertilization and gametophyte development. This conclusion is based on the data on some orchids. Endothelial polyembryony is regarded as a specific case of integumentary polyembryony. According to the authors, no formation of polyembryonic seeds with integumentary or endothelial embryos has ever been reported. Regrettably, some relevant works have been disregarded by the authors which has led to somewhat distorted interpretation of the phenomenon in question.

Classifications of the modes of reproduction in angiosperms have been suggested in a number of works.[13,15,80] Modes of plant reproduction in natural conditions and in tissue culture were incorporated into a single system, though the advisability of such corporation is doubtful.

In summary, a survey of the main publications on general embryology and apomixis demonstrates a variety of opinions among investigators on adventive embryony. The ambiguities which have emerged since Winkler's work still remain.

Chapter 3

STRUCTURAL AND FUNCTIONAL ASPECTS OF NUCELLAR AND INTEGUMENTARY EMBRYONY

I. INTRODUCTION

This section discusses basic problems of the genesis of reproductive structures in species capable of nucellar and integumentary embryony. It is expected to provide a comprehensive embryological description of the phenomenon. Differentiation of initial cells and further development of adventive embryos will be examined in terms of general biology in order to define a theoretical background of adventive embryony in flowering plants.

II. MATERIALS AND METHODS OF RESEARCH

From various groups of flowering plants, 24 species belonging to 12 genera from 6 families were chosen. A study of economic plants whose embryology is unclear (*Citrus limon* cultivars or species of genus *Zanthoxylum*) was undertaken. Material for research was collected from various botanical gardens and natural habitats from 1968 to 1988.

For light microscopy research the material was fixed temporarily (*S. humilis, N. acuminata, B. sempervirens, Opuntia ficus-indica. O. elata, E. macroptera,* and *V. officinale*); fixations were made after artificial pollination daily during the 1st week and once or twice a week, sometimes rarely, until seed maturation. Dissected ovules were fixed in FAA, Navashin, or Carnois mixtures. After fixation, the tissues were embedded in paraphine following standard cytological technique. For staining Heidenhain Haematoxylin with Ruthenium Red or Orange G in Clove oil, Safranin with Alcian Blauw, or Feulgen basic Fuxin with Licht Grun or Erlich Haematoxylin were applied. For histochemistry Unna staining and Procyon dyes following our modification were used.[81] Slides were examined under optical microscope; drawings were prepared using drawing apparatus. Pollen grain fertility was determined by Aceto-Carmin staining following a standard method. Pollen grains were taken from unopened anthers. Well-stained underformed pollen grains were considered as vital. In some cases pollen grains were germinated on 10 to 20% Saccharose in Petri dishes under 20 to 25°C.

For electron microscopy material was fixed in 2, 3 to 5, 5% or Glutaraldehyde in 0.05 M Phosphate buffer (pH 7.2) for 4 h at room temperature or in cold for 6 h; postfixation was in 2% OsO_4 in 0.1 M phosphate buffer (pH 7.8) for 12 h in cold. Further material was washed in several changes of buffer, dehydrated, and stained with 2% Uranil Acetate in 70% Ethanol for 3 h, dehydrated in Alcohol-Aceton ceries, and embedded in Epon 812, Araldite,

Species under study	Locality

Buxaceae

Sarcococca humilis Hort.	Greenhouse of Komarov Botanical Institute, St. Petersburg; Nikitsky Botanical Garden, Yalta; Batumi Botanical Garden
S. hookeriana Bail.	Greenhouses of Kiev State University; Batumi Botanical Garden
S. ruscifolia Stapf.	Greenhouse of Komarov Botanical Institute, St. Petersburg
S. saligna Müll.	Greenhouse of Komarov Botanical Institute, St. Petersburg
Buxus sempervirens L.	Botanical Garden of Ukranian Academy of Sciences, Kiev; Nikitsky Botanical Garden
B. balearica Lam.	Nikitsky Botanical Garden
B. colchica Pojark.	Nikitsky Botanical Garden
Notobuxus acuminata Hutch.	Greenhouse of Komarov Botanical Institute

Cactaceae

Opuntia elata Link et Otto	Greenhouses of Komarov Botanical
O. Ficus-indica (L.) Mill.	Institute

Rutaceae

Poncirus trifoliata Rafin.	Nikitsky Botanical Garden
Citrus limon	
cv. Novogruzinsky	Batumi Botanical Garden
cv. Meier	Batumi Botanical Garden
cv. Ponderosa	Batumi Botanical Garden
	Nikitsky Botanical Garden
C. reticulata Blanko	Batumi Botanical Garden
Fortunella margarita Swingl.	Batumi and Nikitsky Botanical Gardens
Zanthoxylum americanum Mill.	Nikitsky Botanical Garden
Z. ailantifolium Engl.	Nikitsky Botanical Garden
Z. spinifex DC.	Greenhouse of Komarov Botanical Institute
Z. schinifolium Sieb.	Batumi Botanical Garden

Celastraceae

Euonymus macroptera Rupr.	Forest Academy, St. Petersburg
E. planipes Koehne et Koehne	Forest Academy, St. Petersburg
E. Europea L.	Forest Academy, St. Petersburg

Asclepiadaceae

Vincetoxicum officinale Moench	Botanical Garden of Komarov Botanical Institute

Ochnaceae

Ochna atropurpurea DC.	Greenhouse of Komarov Botanical Institute
O. multiflora DC.	Greenhouse of Komarov Botanical Institute

or a mixture of both. Ultrathin sections were cut in ultratom LKB (Sweden) and stained with Uranil Acetate and Lead Citrate following the scheme of Reynolds.[82] Transmission electron microscopes Philips EM-301 and JEM-812 were used for observation.

III. MICROSPOROGENESIS, MICROGAMETOGENESIS, AND MATURE POLLEN GRAINS

Studies of the formation of male generative structures in species with nucellar and integumentary embryony are few in number but they reveal some features of this process. In most cases no special investigations have been undertaken, but the authors indicate the presence of pollen tubes in style tissue, ovary, ovule, or near the embryo sac. These data are available on many species and apparently demonstrate the possibility of the formation of fertile pollen grains. Microsporogenesis and gametogenesis have been studied in more detail in *Citrus limon, Poncirus trifoliata, Smilacina racemosa, Feronia limonia, Zygopetalum mackayi* and other species. The investigators share the view that meiosis and further development of pollen grains in these species are normal and the frequency of formation of fertile pollen grains is high, reaching 70% or more.[83-86]

In some species, however, significant abnormalities are found during meiosis which lead either to collapse of pollen grains or to formation of diploid or aneuploid pollen grains. In these cases the proportion of fertile pollen grains is very small and does not exceed 2 to 3%. This phenomenon is observed in a few species: *Ochna serrulata, Atraphasis frutescens,* and *Alnus rugosa.* Formation of only sterile pollen grains is described for *Euphorbia dulcis, Zeuxine sulcata, Nigritella nigra, Garcinia mangostana, Euonymus japonica, Sarcococca pruniformis, Zanthoxylum americanum, Z. bungei, Z. simulans,* and *Z. planispinum.*[40,108,133,169,170] In our opinion the data on the presence of sterile pollen grains in species should be viewed with caution because its occurrence in one or another species can be confirmed only by special investigations. In other words, it would require studies of this phenomenon undertaken in various regions of geographical distribution of a given species, in a variety of populations, and over different time intervals in order to eliminate a possible effect of unfavorable environmental conditions. No such studies have been performed for any of the species referred to at present as obligate apomicts (i.e., species with sterile pollen grains).

The need for special studies can be illustrated by *E. japonica.* Merzlikina[87] found that nearly 50% of pollen grains of *E. japonica,* cultivated in Crimea, were viable and capable of germination. Meanwhile, pollen grains of this species were described by Copeland[88] as sterile. It is obvious that individuals of this species have both sterile and fertile pollen grains.

TABLE I.
Fertility of Pollen Grains for Species with Nucellar and Integumentary
Embryony (Acetocarmin Stain)

Genus, species	Total number of studied, pollen grains	Stained grains	Unstained grains	% of fertile pollen grains
Opuntia ficus-indica	1484	1266	218	85.30
O. elata	1063	837	226	78.73
Citrus limon Burm.	1456	885	571	60.78
C. limon Novogruzinskii	833	229	604	27.49
C. limon Meier	1757	628	1129	35.74
C. limon Ponderosa	2058	1853	205	90.03
C. reticulata	1656	1535	121	92.69
Fortunella margarita	1719	1658	61	96.45
Zanthoxylum schinifolium	1344	1278	66	95.08
Euonymus macroptera	1441	1424	17	98.82

For our purposes the fertility of mature pollen grains is of utmost importance because it is the resulting factor which reflects preceding events and more
or less permanently predetermines the possibility of double fertilization.

We have found no fertile pollen grains in *S. humilis*, *S. hookeriana*, *S.
saligna* or *S. ruscifolia* (the family Buxaceae), which are characterized by
nucellar embryony. All pollen grains in the two-celled stage appeared to be
degenerating, while they were quite normal in the one-nuclear stage (Scheme
I, parts 1 and 2).* Likewise, trends were observed in both the indoor and
outdoor plants. Normally developed pollen grains of *S. pruniformis* also were
not found.[89-91] However, it should be noted that all these species have been
studied beyond the range of their distribution.

Most other representatives of flowering plants under study which exhibit
nucellar and integumentary embryony are characterized by normal development of pollen grains and their high fertility (Table 1; Scheme III, parts 1 to
7; Scheme V, parts 1 and 2), with the exception of only two varieties of *Citrus
limon* (as a possible consequence of hybrid origin). High fertility of pollen
grains is also observed in *Zanthoxylum schinifolium* and *Z. americanum* (Table
1; Scheme IV, parts 1 to 5). Representatives of the genus have been previously
classified as obligatory apomicts though in the light of new evidence it is not
certainly the case.

Some species of the genus *Euonymus* (the family Celastraceae) with integumentary embryony are characterized by high fertility of pollen grains: in *E.
macroptera* it is equal to 98.8%, in *E. planipes* and *E. europaea* it averages 90
and 80%, respectively. Pollen grains of the species in question have been also
germinated on the 5 to 10% solutions of saccharose at 25 to 28°C. In this case
viable pollen grains amount to 85 to 90% in *E. macroptera* and *E. planipes*, and
to 75 to 80% in *E. europaea*. The data presented by Kordyum[92] and our

* Schemes are presented following Chapter 6.

findings show high fertility of pollen grains of *Vincetoxictun officinale* (the family Asclepiadaceae).

Thus, in most angiosperms which exhibit nucellar and integumentary embryony, processes of microsporogenesis and microgametogenesis follow a normal pattern. Only in a few angiosperms with this apomixis type are no fertile pollen grains found or their percentage is very low.

IV. MEGASPOROGENESIS, MEGAGAMETOGENESIS, AND MATURE EMBRYO SACS

In order to evaluate the significance of the female gametophyte for species with adventive embryony, we have undertaken studies of early stages of development. Information currently available on these aspects is very limited.

Differentiation of megasporocyte follows a normal pattern in all the species under study. The number of megasporocytes per ovule varies from one to three or four depending on the taxon. Megasporocytes and parietal cells are produced by division of the archesporial cells in *Sarcococca, Opuntia, Citrus, Poncirus, Fortunella,* and *Zanthoxylum.* Archesporial cells are the starting points for megasporocytes without division in *Euonymus* and *Vincetoxicum* (Scheme I, parts 6 to 8; Scheme II, part 11; Scheme III, parts 8 to 10; Scheme IV, parts 6 and 7; Scheme V, parts 3 and 4; Scheme VI, parts 1 to 3). We have found no differences between species characterized by adventive embryony and amphimictic angiosperms with respect to these stages. In some species not all of the megasporocytes undergo meiosis and after completion of the process they can be found in different stages. Sometimes meiosis is suppressed during prophase. Among tetrads the earlier formed one usually remain viable, while others gradually degenerate (Scheme III, part 10; Scheme IV, part 10). Such a course of megasporogenesis is common for many other angiosperms characterized by amphimixis.

It should be noted that degeneration of tetrads of megaspores leads to collapse of the ovule in the representatives of some genera under study: *Sarcococca, Citrus,* and *Zanthoxylum* (Scheme IV, part 10; Plate 3, part 3).* It might be a consequence of abnormal meiosis. According to other researchers, in most species with adventive embryony, development of the embryo sacs follows a Polygonum type. It is easily explained since this type of development of the embryo sacs is typical of 87% of angiosperms. This type is known to be rather persistent and deviations from a normal pattern are very rare.

In contrast, our material gives a diverse picture of embryo sac development. Representatives of the genera *Euonymus, Opuntia,* and *Poncirus* show no deviations in the Polygonum-type embryo sac development, like most angiosperms (Scheme II, parts 12 and 13; Scheme V, parts 5 to 7).

On the other hand numerous and diverse deviations are observed in the development of the embryo sacs in *Sarcococca humilis* and *S. hookeriana.*

* Plates are presented following the schemes.

They are manifested in the disruption of polarity when nuclei migrate to opposite cell poles and in their asynchronous divisions in coenocyte phases of development (Scheme I, parts 14 and 15). Such embryo sacs generally collapse prior to maturation. Abnormal cell differentiation of the egg apparatus and the absence of antipodals in the embryo sacs were noted (Scheme I, parts 13, 15). A similar situation was reported for *Trillium camschatcense,* which has no adventive embryony.[81,93] Previous studies and our observations in citrus showed disruption in polarity and position of nuclei in the coenocyte stage of the embryo sac, though normal development was also possible (Scheme III, parts 11 to 15). Moreover, male and female reproductive structures in *Sarcococca* were developed with pronounced time lag; more than 2 months had elapsed since the flowering period when the embryo sacs were formed, while by that time there should have been mature pollen grains.

Another observation concerning the genus *Opuntia* seemed to be important: cells of the egg apparatus in the embryo sacs degenerated even prior to flowering, though some embryo sacs remained normal (Scheme II, parts 1 and 13). This phenomenon occurred both in *O. ficus-indica* and *O. elata,* but in the latter it was much more pronounced. In both species degeneration of the embryo sacs also occurred. A similar tendency was previously noted in *O. dillenii* and in the genus *Citrus.*[94-97] Most frequent abnormalities in the development of female reproductive structures were observed in representatives of the genus *Zanthoxylum,* for which no data on normal megasporogenesis and gametogenesis were reported. As seen from Scheme IV, parts 6 to 10, megasporocytes and tetrads of megaspores usually degenerate.

For other species with adventive embryony information is available on disruptions in the processes of megasporo- and megagametogenesis, which result in a reduced number of viable embryo sacs. In particular, it applies to *Aegle marmelos, O. dillenii,* and some varieties of the genus *Citrus,*[96-100] where nearly 70% (and sometimes even more) of embryo sacs collapse, leading to sterility of the ovules. In some species the egg apparatus can degenerate, as it does in *Opuntia;* sometimes it occurs before flowering, as with *Zeuxine sulcata, Spiranthes cernua,* representatives of the genus *Citrus, Eugenia jambos, Aegle marmelos, Mammillaria tenuis, Zygopetalum mackayi, Eugenia malaccensis, Euphorbia dulcis,* and species of the genus *Zanthoxylum, Aphanamixis polystashya,* and *Euonymus japonica.*[88,100-105]

Some species of angiosperms possessing either adventive embryony or amphimixis can form, apart from monosporic embryo sacs, bisporic and tetrasporic embryo sacs. Their development follows an *Allium*-type in *Allium odorum* and *A. nutans* and *Fritillaria*-type in *Euphorbia dulcis.*[106-108]

A question then arises whether the above deviations are inherent exclusively in taxa with adventive embryony or if similar events may also take place in amphimictic plants. It seems essential to compare the position of the functional megaspore in a tetrad and the pattern of megasporogenesis. In most angiosperms the embryo sacs are known to develop usually from a certain megaspore of a tetrad. Investigations undertaken by Rodkiewicz and co-workers,[109] as well as by other scientists,[110] show that in prophase I of meiosis callose is deposited around the megasporocyte and later around the tetrad. The functional megaspore then becomes free from callose. In addition, megasporocytes of *Gasteria verrucosa*[111,112] and other flowering plants[113] provide an example of polarity in distribution of cell organelles; most are concentrated at the chalazal end of the cell; in this case the chalazal megaspore remains functional. The position of the functional megaspore in the tetrad is predetermined by a number of factors. According to the above histochemical and ultrastructural investigations, prior to and in the course of meiosis megasporocytes undergo a certain preliminary stage of megaspore differentiation before further development starts. Numerous data of light microscope give supporting evidence for certain stability in the position of the viable megaspore in the tetrad. This stability is often a typical feature of a species or even a genus or family as a whole.

Our data show that, in representatives of the genera *Euonymus, Opuntia,* and *Poncirus,* the position of the functional megaspore in the tetrad is usually stable (Figure 1). It is usually the chalazal cell which gives rise to the embryo sac of a Polygonum-type. In embryo sac development there are practically no deviations from a normal pattern. Most achieve maturity, have a typical structure and are viable. Meanwhile, in the genera *Citrus, Zanthoxylum,* and *Sarcococca* the position of the functional megaspore is unstable and any tetrad cell can be this megaspore. Embryo sac development which likewise follows Polygonum-type is in these cases characterized by frequent abnormalities and many embryo sacs collapse prior to maturity. In these genera the proportion of mature and viable embryo sacs does not generally exceed 50% and in some instances it is below 5 to 10%. In these species the entire tetrad often degenerates, leading to sterility of female generative structures even before they begin flowering.

We hold the view that, in the species under study, the unstable position of the functional megaspore in the tetrad adversely affects further development of tetrads and embryo sacs since it results in an increased number of sterile ovules. It appears that not all structural and functional transformations, which would have preceded megaspore differentiation in the case of its stable position, can take place in the case of unstable megaspore position.

Stability of position of the functional megaspore in the tetrad and the resulting pattern of embryo sac development are closely linked; they can be found in a number of amphimictic flowering plants as well (*Quercus, Oenothera,* etc.).

Thus, the deviations observed in megasporogenesis and megagametogenesis in taxa with nucellar and integumentary embryony are not specific features of the given type of apomixis and consequently cannot be regarded as indications of the gametophyte reduction.

Analysis of earlier works in this field has shown that in some species possessing adventive embryony different types of apomixis can coexist. Aposporic and diplosporic embryo sacs are formed in 10% of species which

20 *Apomixis in Angiosperms*

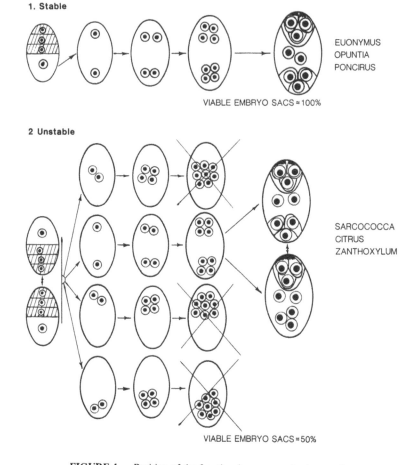

1. Stable

VIABLE EMBRYO SACS = 100%

EUONYMUS
OPUNTIA
PONCIRUS

2 Unstable

VIABLE EMBRYO SACS = 50%

SARCOCOCCA
CITRUS
ZANTHOXYLUM

FIGURE 1. Position of the functional megaspore in the tetrad.

exhibit nucellar and integumentary embryony. Diplospory is observed in *Allium odorum, Eugenia malaccensis, Nigritella nigra, Zeuxine sulcata, Oenothera lamarkiana, Potentilla nepalensis,* and *P. argyrophylla.*[46,47,114-118]

Apospory is found in the genera *Betula, Duschekia, Ribes, Eugenia, Beta, Ochna, Spiranthes, Botriochloa, Poa, Sorghum, Atraphaxis, Malus, Pyrus, Triphasia, Zanthoxylum, Elatostema,* and some other species.[17,119,120,130,133,177-181,194,263,264,269,286,325,336,373]

It should be stressed however that, in most species which have apomictic embryo sacs, embryo sacs of normal pattern with reduced chromosome numbers also develop. The genera *Beta* and *Citrus* can be given as examples.[85,119-121]

It is of notice that unreduced embryo sacs are expected to form in monoembryonic and polyembryonic representatives of the genus *Citrus*.[121,122] These embryo sacs are then capable of fertilization followed by formation of

a triploid embryo and pentaploid endosperm. Seeds which contain these structures are considerably smaller than usual. Among the taxa under investigation,
aposporic embryo sacs are found in *Zanthoxylum americanum* (Scheme IV,
parts 10 to 12).

V. DOUBLE FERTILIZATION

Very few studies consider double fertilization, development of zygotic
embryos, and endosperm formation in species capable of adventive embryony
undertaken. There are only rare instances, when fertilization is reliably described or chromosome numbers of the embryo and endosperm that indicate
their diploid and triploid levels, respectively, are given. This limited number
of species includes *Pachira oleaginea, Momordica charantia*, and others.[106,123,124] In most works the embryos that are formed from egg cells are a
priori assumed to be of sexual origin. However, existing data on the possibility
of development of diploid embryo sacs and the formation of embryos as the
result of parthenogenesis and apogamety require a more careful approach in
defining embryo nature.

Investigators often report the absence of pollen tubes which penetrate into
embryo sacs and degeneration of egg-apparatus cells after flowering. These
features testify that double fertilization does not occur. Similar tendencies have
been encountered in *Cyanella capensis, Smilacina racemosa, Aegle marmelos,
Mangifera indica*, and *Milletia ovalifolia.*[99,125-128]

By contrast, in representatives of other genera (*Citrus, Poncirus, Euonymus,
Opuntia*, etc.) numerous pollen tubes are found in style tissue; some grow,
reaching the embryo sac and even penetrating into one of synergids while
sperms came into contact with egg cell nucleus and polar nuclei, but even in
this case the process of double fertilization is not always completed: fusion of
the gametes may not occur and consequently the egg cell together with the
sperms lying adjacent to its nucleus gradually degenerate.[88,100,129] Double
fertilization is apparently lacking in plants with sterile pollen grains.

Among the species in question, double fertilization takes place in *O. ficus-
indica, O. elata, P. trifoliata, C. limon*, as well as in representatives of the
genus *Euonymus*. Analysis of the data reveals some features of the process.

First, double fertilization is not a regular occurrence and a good many
embryo sacs remain unfertilized. Causes of reduced fertilization have not been
yet discussed in literature. We have compared the fluctuations in growth of
pollen tubes of *E. macroptera* in natural and artificial conditions at controlled
temperature. This has led to the conclusion that low temperatures during the
flowering period have an adverse effect on germination of pollen grains and
growth of pollen tubes. We have observed similar events in four varieties of *C.
limon* cultivated in the northern-most regions of its distribution in Georgia.
They are very sensitive to low temperatures and unfavorable environmental
conditions in general that lead to abnormalities in growth of pollen tubes. There

are also indications that low spring temperatures are harmful to haploid embryo sac formation in apple trees which are replaced by diploid aposporic embryo sacs.[130]

Second, we have recorded abnormalities in the course of double fertilization itself. In particular, one of its stages is omitted and as a consequence fertilization of either polar nuclei or egg cell is achieved (Scheme III, parts 17 and 18). In the latter case the ovule collapses due to the lack of endosperm for a developing embryo. In the first case the ovule may remain functional, provided the nucellar embryos are formed. Disruptions of double fertilization found in *C. limon* and *E. macroptera* as well as in *Poa pratensis*[131] are not isolated instances that are observed only in species capable of adventive embryony. They are not uncommon in a number of amphimictic species: *Dactylis glomerata, Phleum pratense,* etc.[132]

Third, double fertilization has a varying degree of occurrence in different taxa. It is relatively frequent in *O. ficus-indica, E. europaea,* and *Poncirus trifoliata;* it is less common in *O. elata, E. macroptera, E. planipes, C. limon,* and *C. reticulata;* it occurs very rarely, if present at all, in very few genera: *Sarcococca, Zanthoxylum,* and *Ochna.* Nevertheless, unfavorable environmental conditions seem to be not the sole, though primary, cause of irregular double fertilization. Hybridization may directly affect the course of embryological processes in much the same manner.

VI. ENDOSPERMOGENESIS

The endosperm in species with sterile pollen grains is of apomictic origin. The mode of endosperm formation in the case of fertile pollen grains is not immediately clear. Of particular interest are works where ploidy of the endosperm is given or double fertilization is described. These investigations show that in species capable of adventive embryony the endosperm can originate in two ways: (1) from fertilization of polar nuclei of the central cell by one sperm or (2) via apomixis, without fusion of this nuclei with the sperm. The first mode is observed, for example, in *Pachira oleaginea, Momordica charantia, Nothoscordum fragrans,* and in the genus *Citrus,*[86,106,123,124] while the second one is reported in particular in *Smilacina racemosa, Atraphaxis frutescens,* and *Scopolia carneolica.*[126,133-135] Also, there are indications that both types of endosperm formation may coexist in the same species. Nygren[3] justifiably has observed that this issue is rather complicated and further studies are needed for its clarification.

Development of apomictic endosperms in flowering plants has been examined by Kandelaki.[136] This work is based on the observations of endosperm development of wheat in comparison with some other taxa (*Antennaria, Alchemilla, Chondrilla, Taraxacum, Ranunculus*). Kandelaki states that apomictic endosperm can arise from division of each polar nucleus formed by their fusion, with the first divisions being characterized as amitotic. Later, the pattern of divisions is normalized and they usually become mitotic.

We have found that in the species under study endosperm can be formed either by amphimixis or apomixis. In *Opuntia ficus-indica, Citrus limon,* and *Poncirus trifoliata* the endosperm originates as the result of fertilization. In *Euonymus macroptera* and probably in *O. elata* the endospem might be of sexual or apomictic origin; in the latter case it originates from fusion of polar nuclei and their subsequent division. Divisions of the endosperm nuclei without signs of fertilization in the egg apparatus are observed in *E. macroptera* (Scheme V, part 8) and apomictic endosperm has 2n = 36.[71,72] Only the apomictic endosperm is typical for species of the genus *Sarcococca* in the presence of sterile pollen grains. However, in *S. humilis,* unlike *E. macroptera,* the apomictic endosperm can arise not only by fusion of polar nuclei; the process can also involve nuclei of undeveloped antipodals (Scheme I, parts 16 to 20): fusion of three to five nuclei, including two polar nuclei, can be observed in the central cell.[73]

There are other features of the apomictic endosperm. It is known that there is a close correlation between the rates of embryo and endosperm development in typical amphimicts.[137] This relationship is not established in species with adventive embryony. We cannot attribute this just to the apomictic nature of the endosperm, but its possible role cannot be dismissed. In *O. elata,* where only nucellar embryos primarily develop, cell formation in the endosperm starts 3 weeks later than in *O. ficus-indica,* which has sexual embryos. In many species under study the onset of cell formation in the endosperm does not depend on the degree of embryo differentiation. Thus, within the same species but in different ovules one can find large globular embryos in the presence of nuclear endosperm and newly developing adventive embryos, in the presence of cellular endosperm, e.g., in the genus *Euonymus* (Plate 4, parts 10 to 12), and in genera *Opuntia, Citrus, Zanthoxylum* and *Vincetoxicum* (Plate 5). Moreover, in species capable of adventive embryony, nucellar and integumentary embryos do not arise simultaneously and their early development can be a lasting process; some embryos might be in the stage of organogenesis, although others consist only of several cells. Stages of endosperm development at the time of early formation of these embryos are different because it occurs in the same embryo sac (Plate 1, part 4; Plate 2, parts 6 and 7; Plate 3, parts 11 to 13; Plate 4, parts 10 to 13).

Now we turn to the role played by the endosperm in embryo development. Numerous investigations show that in natural conditions embryos are not capable of normal development in the absence of the endosperm. Only some specialized taxa (Orchidaceae, Podestemonaceae, etc.) can be given as exceptions, where the endosperm function is replaced by the function of other structures. This also applies to embryo development from nucellus or integument in experiment. In the absence of endosperm they inevitably collapse in the early developmental stages.[46,47,138,139] A more recent work provides supporting evidence for this view.[77] The authors show that nucellar embryos might start to develop in unpollinated ovules in the absence of endosperm. However, their formation is stopped in either a globular or torpedo-like stage and the seeds with such embryos are not able to germinate in natural conditions. These observations are consistent with our data (Scheme III, parts 16 and 19; Scheme V, parts 10 to 12; Plate 4, parts 4, 5, 12, and etc.). However, an opposite view that we do not share is sometimes expressed.[140,141]

VII. INITIAL CELLS OF NUCELLAR AND INTEGUMENTARY EMBRYOS — EMBRYOCYTES

Nucellus and integument cells that can give rise to adventive embryos have been called initial cells or embryocytes. The latter term was suggested by Yakovlev (personal communication). Both of these terms are used in our work. Embryocytes are unique structures which are characteristic only of the ovules of flowering plants. Their origin, structure, position within the tissue, and other aspects are the focus of many investigations.

The appearance of initial cells and adventive embryos within the nucellus or integument tissue often seems an unusual phenomenon. Examination of the fine structure of these cells might be helpful in clarifying this issue. It has been known since the work of Strasburger[29] that these cells have large nuclei, optically dense cytoplasm, and are intensively stained by various dyes. Previous studies made at the light microscopy level have not essentially improved our understanding of embryocyte structure.

Starch grains were found in the initial cells of *Mangifera indica*.[142] Later, starch was observed in initial cells of other species, though its presence was not permanent. Formation of a thickened cell wall was noted in some species. Many investigators regarded initial cells of adventive embryos as identical to the zygote.

Only recently have ultrastructural studies of embryocytes been undertaken for *Euonymus macroptera, Sarcococca humilis,* and *Poncirus trifoliata*[143-147] and for some *Citrus* cultivars[148] Embryocytes and other nucellus and integument cells have been under study in the stage shortly before nucellar embryo formation. Figure 2 schematically presents the ovule of *S. humilis* at the time of embryocyte differentiation. Also shown are central sections of its major cell types, indicating the number and distribution of cell organelles. Figure 3 illustrates mitosis of the embryocyte and early developmental stages of nucellar embryos. In constructing these two figures electron microscopy composition of the pictures were used. Figure 4 gives ultrastructural characteristics of embryocytes and other cells of the ovule, as well as data on embryocyte mitosis and nucellar embryo structure.

Embryocytes are relatively large and have a more dense cytoplasm in comparison with usual nucellus or integument cells (Figures 2, part F, and 4; Scheme III, parts 16, 19 and 21; Plate 1, parts 1 and 5; Plate 3, part 6; Plate 4, parts 3 and 6). The cell nucleus is large and irregular in shape, frequently with deep invaginations, and it takes up most of the cell. There are one or several nucleoli, which have "vacuoles" and consist of granular and fibrillar components.

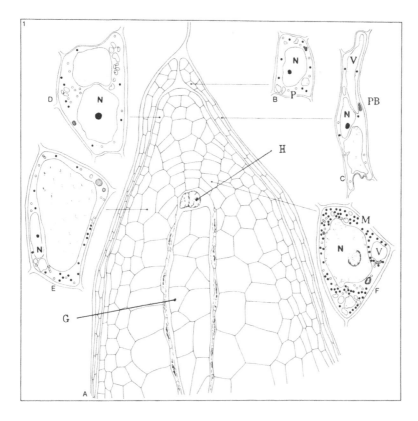

FIGURE 2. Ovule of *Sarcococca humilis*. (A) Cell pattern of ovule (3 months after flowering), (B and C) cells of inner integument, (D) nucellar epidermis, (E) cell from lateral nucellar part, (F) cell from micropillar nucellar part — initial cell of nucellar embryo (embryocyte), (G) endosperm, (H) degenerating egg apparatus. Consult list on page 85 for other abbreviations.

Additional nucleolus-like structures are often observed. Pores in the nuclear membrane are well pronounced and numerous (Plate 1, part 6; Plate 4, parts 6 and 7). Polysomes and free ribosomes are abundant and distributed quite evenly throughout the cytoplasm. Rough and smooth endoplasmic reticulum are represented by isolated cisternae. Vacuoles differ in size, have electron-dense inclusions, and are located more or less evenly on the cell periphery. Dictyosomes consist of five or more cisternae and produce few vesicles. Numerous mitochondria are of varying shape and size, are uniformly spread throughout the cell, and some concentrated near the nucleus. They can be spherical, cup-like, or elongated and dumbbell shaped. All mitochondria contain swollen cristae. Plastids, like mitochondria, are abundant and vary in shape and size: from small and oval to elongated and dumbbell shaped. Some plastids contain starch grains. Cup-like plastids are sometimes found. Electron densities of mitochondrial matrix and plastid stroma are approximately equal. Plasma

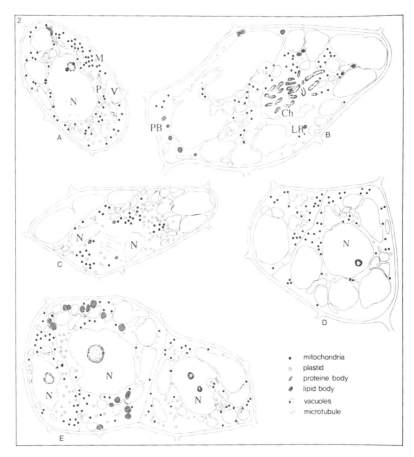

FIGURE 3.   Mitosis of embryocytes and two- to three-celled nucellar embryos. (A) Prophase, (B) metaphase, (C) telophase, (D) two-celled nucellar embryo, (E) three-celled nucellar embryo. Consult list on page 85 for other abbreviations.

membrane is wavy; it forms distinct invaginations toward the cytoplasm. Polarity of the cell organelle array is not pronounced (Plate 1, parts 6 to 10; Plate 4, parts 6 to 9).

The cell wall is thickened. In *Sarcococca humilis* at the developmental stage in question the cell wall has plasmodesmata (Plate 1, parts 5 and 8), which later disappear (Plate 1, parts 11, 15, and 17 to 21). In *Euonymus macroptera* we have not found plasmodesmata in the cell wall of the embryocyte; a concentration of electron-dense substance was observed. In the embryocytes of representatives of the genus *Citrus* plasmodesmata are present only at early stages of their differentiation and then disappear.[148] In general, reorganization processes occurring in the cell wall structure of the embryocyte are still inadequately understood. However, it can be assumed that the loss of plasmodesmata is one of the typical features of the embryocyte differentiation and development of adventive embryos. The time of this process occurrence is probably typical of the species.

The above ultrastructural characteristic of embryocytes, their comparison with other cells of the nucellus or integument which are incapable of embryo formation, and the few published data[148] show their most characteristic features. Embryocytes have very large nuclei, often of irregular shape, with large nucleoli which consist of granular and fibrillar components; abundant free ribosomes and polysomes; and a great number of mitochondria and plastids of varying shape and size, whose structures testify to their high functional activity. The above cell organelles are major indications of metabolic activity and energy potential of the cell which in this case is very high. This contention is confirmed by intensive staining of embryocytes following Unna's reaction and by Procion dyes to identify proteins and carbohydrates (Plate 4, parts 15 and 16). Moreover, we have found that embryocytes differ in their metabolic processes from other cells of the tissue. In embryocytes of *Citrus* in particular starch is synthesized most intensively and there are very few, if any, lipid bodies, although in other nucellus cells synthesis of lipids is predominant. In embryocytes of *Euonymus macroptera* synthesis of carbohydrates (starch) dominates while other integument cells synthesize proteins in large quantities (Plate 4, part 10). Other investigators note a high concentration of ascorbic acid in the nucellus cells of *Citrus* which are adjacent to the embryo sac and are capable of producing nucellar embryos.[149] The processes of synthesis in embryocytes must be attributed to changes in the cell wall structure.

Plasmodesmata are known to play an appreciable role in symplastic transport: they carry directly certain substances and signals from cell to cell.[150,151] Breakdown of plasmodesmata leads to physiological isolation of a cell. In this case substance transport between cells becomes apoplastic, while in plant tissues as a whole it is symplastic. Many questions related to the structure and specific functioning of plasmodesmata as well as qualitative differences between symplastic and apoplastic transport remain unresolved.

However, the available data indicate that the loss of plasmodesmata has serious implications. Substances incoming to the embryocyte become more selective. Limited contact with surrounding cells is an essential feature for embryocyte differentiation and new organism (embryo) origination.

Changes in the cell wall structure of embryocytes and adventive embryos are manifested morphologically in their relationships with surrounding cells. This issue was first addressed by Yakovlev.[65,66] He supposed that initial cells should be isolated from surrounding cells; as a result nucellar tissue disintegrated entirely. Our observations also indicate that during embryocyte differentiation and adventive embryo development degenerating cells emerge in the vicinity of these structures (Plate 3, part 6; Plate 4, parts 3, 4, 10, 15, and 16). The appearance of degenerating cells seems to be a consequence of limited contacts between the embryocyte and adjacent cells of nucellus or integument. Normal development of embryological structures are accompanied by a similar process. In particular, degeneration of nucellus cells adjacent to the embryo sac is common, as known plasmodesmata are absent in the embryo sac wall except for its chalazal part.

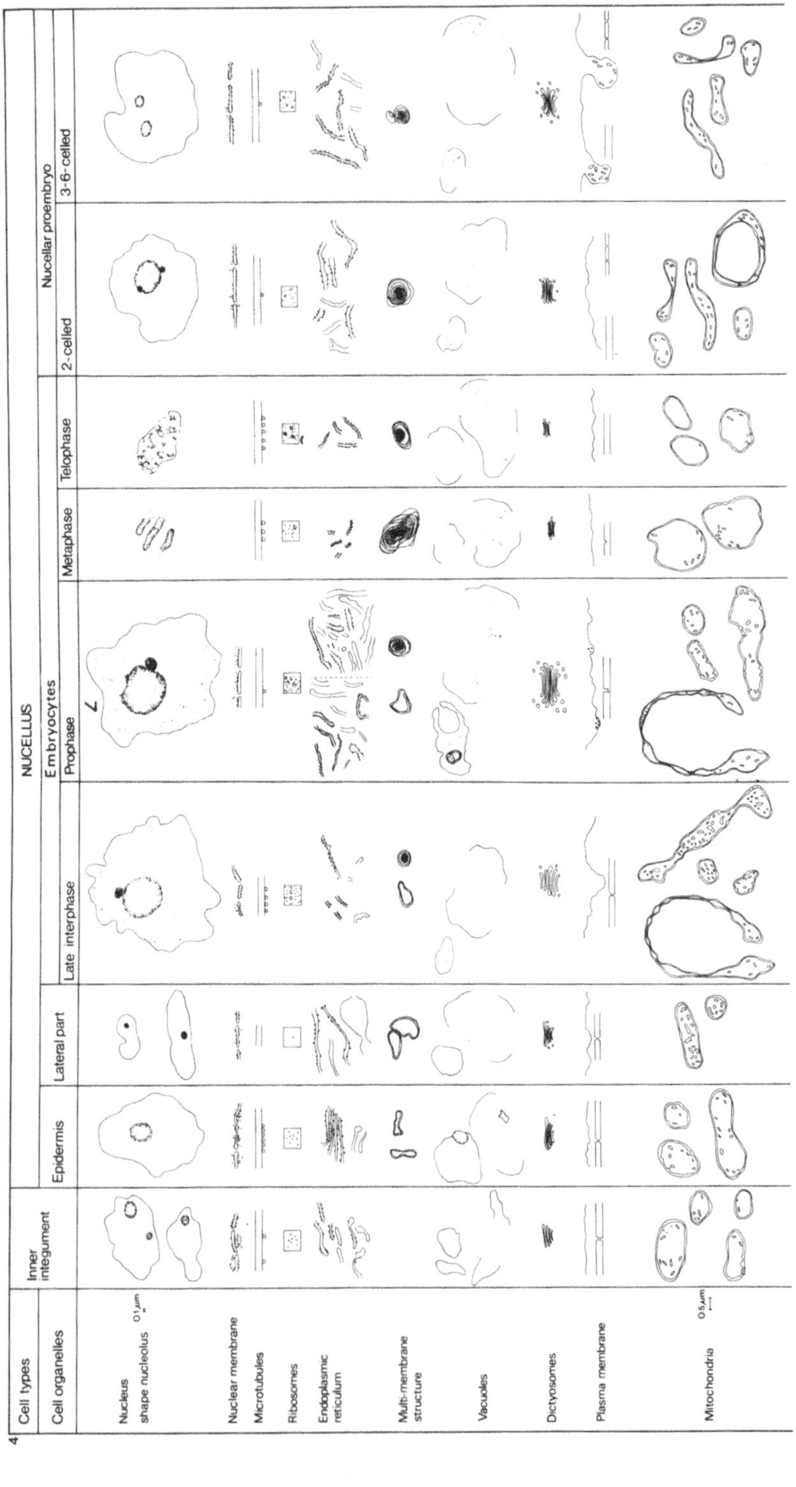
4
Cell types
Cell organelles
Inner integument
NUCELLUS
Epidermis
Lateral part
E mbryocytes
Late interphase
Prophase
Metaphase
Telophase
Nucellar proembryo
2-celled
3-6-celled
Nucleus shape nucleolus
0.1 μm
Nuclear membrane
Microtubules
Ribosomes
Endoplasmic reticulum
Multi-membrane structure
Vacuoles
Dictyosomes
Plasma membrane
Mitochondria
0.5 μm

29

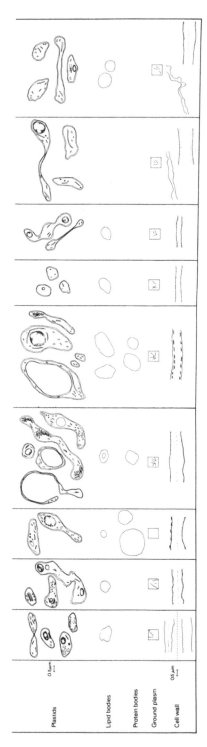

FIGURE 4. Ultrastructural characteristics of embryocytes, two- to three-celled nucellar embryos, inner integument, and nucellus cells.

VIII. MITOSIS OF EMBRYOCYTES IN *SARCOCOCCA HUMILIS*

We have obtained data on the ultrastructure of a protoplast during consecutive stages of mitosis (prophase, metaphase, and telophase) resulting in the production of a nucellar embryo (Figures 3 and 4). A comparison of the results of this study shows that certain transformations occur in a protoplast during this process. They are reflected in structural changes of plastids and mitochondria in particular, in varying number of free ribosomes, polysomes, and endoplasmic reticulum, and in the changing activity of dictyosomes. Apart from that in the course of division embryocytes undergo maximal vacuolization at metaphase and lose plasmodesmata (in *S. humilis*) connecting embryocytes earlier with neighboring cells. One should also note an even distribution of cell organelles at prophase and consequently the absence of cell polarization (Plate 1, parts 11 to 15) which will be shown to play an essential role in adventive embryogenesis. A question then arises whether these protoplast transformations during mitosis of the initial cells are peculiar features or if they are usual for mitosis of any somatic cell and what they have in common in this respect with meiosis. Regrettably, no studies of ultrastructural transformations of cytoplasm during mitosis of nucellar cells have been performed so far and data on other cell types are also scanty. Ultrastructural transformations of cell organelles which occur during meiosis of micro- and megasporocytes have been treated in more detail.[152-154] Similar data have been obtained by Gabaraeva[155-157] on *Psilotum*.

A comparison of our data on ultrastructural transformations of protoplast during mitosis of *S. humilis* embryocytes with published data on megasporocyte meiosis of flowering plants shows many common features. In particular in both types of division one can note structural transformations of mitochondria and plastids and changes in density of ribosomes and polysomes. Strong marked vacuolization of cells is observed in both cases. Cell vacuolization during meiosis is usually regarded as the main process providing for alternation of the old sporophytic cytoplasm by a new gametophytic one. Another essential factor is the loss of plasmodesmata observed during mitosis in *Sarcococca* and probably at some earlier stage in other species (*Poncirus, Euonynus*). In these two types of division thickening of the cell wall takes place. Apart from a common set of varying characters, some differences are revealed for *Psilotum nudum*.

Common features of transformations in protoplasts of embryocytes and megasporocytes can be easily explained. Initial cells of adventive embryos and megasporocytes are in this case the cells which give rise to reproductive structures, i.e., adventive embryos and gametophytes, respectively. The protoplast transformations are definitely controlled by the nucleus, which determines qualitative and quantitative variations of the cell metabolism. These processes take place both in megasporogenesis which is typical of all angiosperms and in the case of adventive embryony, although they occur at different stages of ontogenesis and are linked with meiosis and mitosis, respectively. It can also be stated that genome reduction during meiosis is not necessary for ultrastructural reconstruction of megasporocyte protoplast because in mitosis of the embryocyte similar processes of protoplast reorganization might occur in the case of adventive embryony. Reorganization of the protoplast seems to be the mechanism responsible for the emergence of a new generation. This mechanism does not depend on the mode of reproduction of a given generation; the latter can be a result of meiosis and amphimixis or develop by apomixis and mitosis in the absence of meiosis.

According to published data the cytoplasm is an essential component of cell differentiation. The capacity of the cytoplasm to affect gene expression which brings about cell differentiation during embryogenesis has been demonstrated in recent investigations. The role played by the cytoplasm in the formation of plant apomictic embryos still has not been under study. Our notions on the role of megasporocyte meiosis and embryocyte mitosis are not traditional; indirect evidence is needed to discuss them.

It would be reasonable to compare initial cells of adventive embryos with an archesporial cell and with megasporocytes, i.e., with those generative cells that are differentiated as single elements at early stages of ovule development. Their external resemblance deserves notice: sporogenic cells and embryocytes are larger than surrounding cells and their nuclei are particularly large. Ultrastructural studies of angiosperms and some primitive vascular plants (i.e., *Psilotum*) show that archesporial cells and megasporocytes are rich in cell organelles, such as mitochondria and plastids, and they contain a great deal of free ribosomes and polysomes. These cells are not isolated from other ones at their early developmental stages because plasmodesmata are at first the components of the cell wall,[110,156,158,159] while plasmodesmata disappear later. Ultrastructural data demonstrate a certain affinity of initial cells to megasporocytes in the protoplast structure.

An essential difference should be noted, however: the polarization of megasporocytes, and the composition and quantity of cytoplasmic components at the poles are not identical,[112,160] while embryocytes do not show any polarity in the cell organelle pattern.

Embryocytes are similar to the zygote in their capacity for embryo production. It is common knowledge that after fertilization density of cytoplasm and vacuole reduction in zygote occur. The number of mitochondria and plastids in the cytoplasm is growing and organelles are increasing in size; they become more diverse in shape and their structure is more complex; this is accompanied by final rebuilding of the cell wall. Thus, the zygote becomes completely surrounded by the cell wall, plasmodesmata disappear, and the substance entrance becomes more selective. The differentiation process of the zygote has many common features with that of embryocytes: they lie in what classifies these cells as being differentiated and isolated from surrounding cells and having high metabolic activity. However, their comparison also shows some differences: the zygote is the major element in the complex system of the embryo sac and is a polarized structure, while embryocytes are not.

Therefore, similarities in the general reorganization during embryocyte, megasporocyte, and zygote development confirm the idea of universal principles used to differentiate embryological structures in the case of amphimixis and adventive embryony.

There are other aspects connected with embryocyte formation. In many investigations attempts have been made to identify the factors that might promote the emergence of initial cells. Controversial views have been expressed on this point. Most researchers believe that pollination is a requisite for initial cell formation and early development of adventive embryos. This opinion was put forward in studies of the genera *Citrus, Mangifera indica, Smilacina racemosa,* and *Vincetoxicum officinale*.[45,149,161-167] However, an opposite view has been advocated which denies the role of pollination in the process.[106,126,168] It is based on the data which indicate that adventive embryos can develop in plants with sterile pollen grains: *Euphorbia dulcis, Alnus rugosa,* and *Atraphaxis frutescens*.[40,108,133,169,170] There are data on the formation of viable seeds with adventive embryos in the absence of male individuals: *Garcinia mangostana* and *Alchornea ilicifolia*.[28,170,171] One of the latest and most detailed works on adventive embryony points out that initial cells of nucellar embryos in representatives of the genus *Citrus* are formed before flowering, this process being observed both in pollinated and unpollinated ovules.[78] Other investigators believe that formation of initial cells and further production of adventive embryos from them are closely connected with the presence of the zygote embryo.[44,164,173,174]

The idea that adventive embryos could arise as the result of mechanical influences on the ovule (e.g., damages made by insects) was entertained by Haberlandt.[134,135,175] In his studies of apomixis forms he advocated the hormone theory of apomixis origin. Later, investigations of *Poaceae* have shown that aposporic embryo sacs and nucellar embryos can develop without external forcing; these studies concern *Agropyron, Sorghum bicolor, Botriochloa ischaemum, Pennisetum mezianlun, Centhrus ciliaris, and Poa pratensis*.[141,176-182] The present studies on *Poaceae* and earlier test experiments on other plants have not revealed any evidence (excluding Nishimura's data[183,184]) for the hormone theory put forward by Haberlandt. In fact, disruption of nucellus cells is a regular developmental stage in both apomictic and amphimictic representatives of angiosperms (e.g., they are disrupted during embryo sac development).

Experimental studies have also been performed to identify factors that might stimulate the formation of initial cells and adventive embryos.[46,47,138,139,173] The authors agree in general that formation of initial cells and early development of adventive embryo are irrespective of pollination, fertilization, and the presence of zygote embryos. Fagerlind,[19] who investigated some representatives of the genera *Hosta,* assumed that the ovule could be prepared to produce adventive embryos by applying physiologically active substances; however, he pointed out that the origin of the factors stimulating apomictic endosperm development was unknown. Negative results have been obtained by other researchers who tried to induce nucellar embryony in representatives of the genera *Citrus* by some hormones.[142]

The results obtained by American scientists are noteworthy.[86] They studied varieties of *Citrus* and made some conclusions. First, pollination stimulates neither seed development nor adventive embryogenesis. Second, adventive embryo development is closely connected with the presence of endosperm, which is mostly formed due to fertilization and is triploid. Sometimes it might be pentaploid, which can probably be the result of the formation of unreduced embryo sacs. Third, the onset of adventive embryo development does not depend on that of zygote division.

Further studies in this field have been undertaken by Japanese scientists.[185,186] The authors suggested a hypothetical mechanism of nucellar embryo origin; this scheme is based on their own data on polyembryonic and monoembryonic *Citrus* progeny *in vivo* and *in vitro* and on relevant published data. They believe that differentiation of initial cells of nucellar embryos (which they call primordial cells) is controlled by the genetic factor (gene of polyembryony); these cells begin to divide under the stimulating effect of fertilized egg cell division; further development of nucellar embryos is associated, according to the authors, with the nutrition function of the endosperm. None of the above viewpoints outweighs others.

In our study of *C. limon* and *C. reticulata*, we have found, as mentioned earlier, degenerating ovules with numerous initial cells of adventive embryos in the nucellus. Cells of the egg apparatus and nonfused polar nuclei remain unfertilized and show distinct signs of degeneration. In *Sarcococca humilis* and *S. hookeriana*, embryocyte differentiation is observed in the absence of fertile pollen grains. In representatives of the genera *Zanthoxylum* and *Vincetoxicum* no pollen tubes are found in the vicinity of embryo sacs. We believe these observations show that embryocyte differentiation does not depend on pollination, fertilization, or the presence of the embryo from zygote. Moreover, first divisions of initial cells are found to occur in the absence of sexual embryo, when the zygote or egg cell is degenerating. In *C. limon* and *C. reticulata*, first division of the initial cell can be observed in the absence of endosperm in the degenerating ovules (Scheme III, parts 16 and 20).

In light of these negative results with respect to pollination and fertilization, we suggest our interpretation of these processes. Differentiation of initial cells and early development of nucellar embryos are an implementation of the genetic program for the development of nucellus cells; this program might be realized due to the activity of certain genes which are inactive in amphimictics. This implies that adventive embryony is linked to peculiarities in the functioning of the plant genome.

The time of appearance and location of embryocytes are also important. In most angiosperms capable of adventive embryony, initial cells become recognizable from other nucellus or integument cells in the case of the mature embryo sac during flowering period or shortly after its completion. To these plants we refer representatives of the genera *Citrus, Poncirus, Opuntia, Vincetoxicum,* and *Sarcococca.* Only in some species can initial cells be singled out at earlier developmental stages, i.e., before flowering. In *Zygopetalum mackayi* initial cells are observed at the tetrad stage of megaspores.[187] It can be inferred from analysis of drawings of *Eugenia jambos* given by van der Piji[48] that differentiation of these cells takes place when embryo sacs are still at the two-nuclear stage. In *Nothoscordum fragrans* and *Euphorbia dulcis* initial cells are clearly seen at the four-nuclear stage of the embryo sac.[102,188] It deserves notice that initial cells exist in the nucellus or integument for a long interval. In many species they can be observed in developing seeds with adventive embryos varying in shape and stage of development (Scheme III, part 19; Plate 4, parts 2 to 5 and 10 to 13; Plate 5, parts 1 to 4).

The position of embryocytes within the tissue can vary. Since Strasburger's work,[29] these single cells are known to often be found in the micropillar part of the nucellus in the case of nucellar embryony (genera *Citrus, Poncirus, Mangifera,* and many others). Our observations on the genus *Sarcococca* which has crassinucellate ovules (Scheme I, parts 3 to 5) indicate that practically all cells of the micropillar part of the nucellus can become embryocytes. Any of these cells can start division, forming the nucellar embryos. As a consequence a great many embryos adjacent to each other appear (Plate 1, parts 1 to 4). Single initial cells located in different parts of nucellus are observed in the genera *Citrus* and *Poncirus* (Scheme III, parts 16, 19, and 21; Plate 3, part 6). According to some investigators initials of nucellar embryos are located in lateral and chalazal parts of this tissue.[77,78] Initials of integumentary embryos are most frequently found in the micropillar part of the integument.[129]

A review of relevant publications shows that initials of adventive embryos might arise not only in the tissue of nucellus or integument formed before flowering. In some species a special tissue is formed; it is denoted by various terms. Possible formation of this tissue was first reported in *Hiptage madablota.*[189,190] Kapanadze has also described such a formation in *Citrus* which he called "tapecial tissue". It consists of two to three layers of cell adjacent to the embryo sac and its size reaches to half a length of the embryo sac. A similar phenomenon is observed in representatives of the genera *Chimonanthus* and *Calycanthus.*[103,104,191,192]

Our studies of *Opuntia elata* show that after flowering the meristematic nucellar tissue is formed, enveloping the embryo sac like a capsule. Separate cells of this particular tissue become embryocytes and give rise to nucellar embryos (Scheme II, parts 20 to 24; Plate 2, parts 4 to 7). A similar occurrence was later found in the genera *Citrus, Poncirus* (Scheme III, parts 20 to 22; Plate 3, parts 8 and 9), and *Zanthoxylum* (Scheme IV, part 13; Plate 3, parts 13 and 14) which are characterized, like the genus *Opuntia,* by nucellar embryony and a crassinucellate type of ovule.[193,194]

Formation of such a tissue in species with a tenuinucillate type of ovule is also observed in the genera *Euonymus, Vincetoxicum,* and *Ochna* in the case of integumentary embryony (Plate 4, parts 1 to 3, 5, and 10; Plate 5, parts 2 to 4; Scheme VII, part 1). It should be noted that in *E. macroptera* the formation of meristematic integumental tissue after flowering occurs both in the micropillar and chalazal parts of the inner integument. This leads to frequent development of embryos in the seed which are formed at opposite poles and grow toward each other (Scheme V, parts 41 to 42; Plate 4, parts 12 and 14).

Whether adventive embryos always develop from single initial cells is a matter of debate. Some investigators suggest the existence of two modes of nucellar embryo formation: from single initial cells similar to the zygote and from a group of such cells combined into an embryonic cell complex.[181,182,239]

We believe that nucellar and integumentary embryos in natural conditions originate from a single initial cell — embryocyte. This view is based on our results with respect to the genera *Sarcococca* (Plate 1), *Opuntia* (Scheme II, parts 14 to 16), *Citrus* and *Poncirus* (Scheme III, parts 19 to 22; Plate 3, parts 6,7, and 10), *Euonymus* (Plate 4, parts 3, 4, 10, 16, and 17), *Vincetoxicum* (Plate 5, part 1), and on principles of embryocyte differentiation.

Initial cells do not immediately start to divide and form adventive embryos. There is an interval between their appearance and beginning of division which is provisionally called a period of relative dormancy. This period can vary in time in different species and it lasts from several days in hybrids *Potentilla nepalensis* × *P. splendens*[118] to several weeks in *E. macroptera*[72] and in *Citrus*[44,78] or even months in *Zanthoxylurn americanum*[194] and in *Zygopetalum mackayi*.[187] This period probably plays an essential role in changing the cell metabolism. No data are found on this issue in the literature.

During the preparatory stage of embryocytes to the embryo formation, significant changes are observed in the embryo sac. In representatives of the genera *Sarcococca, Zanthoxylum,* and *Ochna* the egg apparatus cells degenerate and the endosperm starts to develop (Figure 2; Scheme I, parts 16 and 17; Scheme IV, parts 13 and 14; Scheme VII, parts 1 and 2). Degeneration of the egg apparatus is also possible in the genera *Euonymus, Opuntia, Citrus,* and *Vincetoxicum,* which is accompanied by endosperm development (Plate 2, part 4; Plate 3, parts 8 and 9; Plate 4, part 1). Thus, in species capable of adventive embryony in many cases only the central cell (with developing endosperm) remains viable among all cells of the embryo sac.

IX. DEVELOPMENT OF SEXUAL AND ASEXUAL EMBRYOS

Embryo development from the nucellus or integument has not been thoroughly studied in any representative of flowering plants, and ultrastructural data were lacking. We have attempted to fill in the gap. As noted earlier, in *S. humilis* all cells of the micropillar part of the nucellus become embryocytes. In the beginning of embryogenesis in the micropillar part of the nucellus, numerous embryos arise that lie close to each other and are located both near the embryo sac and at some distance from it. During this period the development of embryos is concurrent; some of them degenerate. Initial cells that do not start to divide can also degenerate (Plate 1, parts 2, 3, and 20).

Both cells of a two-celled nucellar embryo produced by the initial cell mitosis are very similar in ultrastructure. They do not differ in size and have large nuclei, often of irregular shape, with one or several nucleoli; vacuoles are located along the cell periphery. The cell walls of dividing daughter cells are very thin and have numerous plasmodesmata, although the external cell wall of a mother initial cell is thickened where plasmodesmata are unclear, rare, and probably reduced to nonfunction (Plate 1, parts 15 to 19). Both cells of this embryo do not differ in the number of mitochondria and plastids (Figure 3). All cell organelles are evenly distributed between them and have similar ultrastructural characteristics. Free ribosomes and polysomes are abundant, though in less quantity than in initial cells. Endoplasmic reticulum is represented by separate cisternae. Vacuoles are not usually large with electron-dense inclusions. Mitochondria do not vary widely in shape and size; most are oval or elongated, the system of cristae is undeveloped, and the matrix is translucent. Plastids, like mitochondria, are uniform in shape and size; they are small, oval, or elongated, the stroma is dense, and thylakoids are scarce. Lipid and proteins bodies are not present (Plate 1, parts 15 to 19; Plate 3, part 10; Figure 3).

Ultrastructural investigations allow to state that the structure arising due to the mitotic division of the embryocyte is a two-celled embryo. This contention is derived from the observation that the external cell wall or the cell wall of the mother initial cell remains thickened and that it loses plasmodesmata (Plate 1, parts 15 to 22). Thus, this new two-celled formation is isolated in contrast to the usual meristematic cells which are in close contact with each other and with adjacent nucellus cells. Further development of the formation into the embryo confirms the above conclusion.

At some later developmental stage, (three- to six-celled nucellar embryos differ slightly from two-celled ones in ultrastructure of their cells (Figure 4; Plate 1, parts 19 to 22). One can only note somewhat higher activity of dyctiosomes and increased plasmalemma invaginations, for which intensive growth of cells and cell vacuolization are probably responsible. Lipid bodies appear. There is a distinct tendency for thickening of the external cell wall; not even remains of plasmodesmata present earlier in the embryocyte cell wall are found (Plate 1, parts 20 and 21; Plate 3, part 10). Consequently, we propose to reject the existing concept of adventive embryos as "offshoots" or "buds" of the nucellus or integument tissue.

It is of note that nucellar embryos at the beginning of their development are lacking distinct polarity, like the initial cells which produce them. The absence of polarity and unregulated growth of embryos at early stages can be illustrated by Plate 1, part 20. Signs of polarity and intensive cell division are first observed in these nucellar embryos which are located in the immediate vicinity of the embryo sac. It also deserves notice that intensive cell division of nucellar embryos results in their somewhat simplified ultrastructure. It is possible that this simplification of cytoplasm ultrastructure is essential for the beginning of embryo differentiation and organogenesis.

Most sexual embryos are known to be characterized by unequal division of the zygote: cells "ca" and "cb" are of different size. In addition, differentiation of zygotic embryos starts from very early stages of development.[195-198] Disruption of this process might result in embryo collapse.[199] In all types of embryogenesis with respect to zygotic embryos the key role is attached to their early developmental stages when both the direction and sequence of divisions are well ordered.[1,200,201] These particular stages allow to trace the origin of cells, the number of previous generations undergone, and the position of derivative cells. This makes it possible to predetermine their contribution in further embryo development.

Investigators agree in general that adventive embryos at early stages of development are multicelled formations without strictly defined shape. However, it holds true irrespective of species in question. First division of embryocyte might take place in different directions (e.g., *Euphorbia dulcis*);[102] there is no definite sequence in divisions of proembryo cells (*Zeuxine*);[202] there is no strict gradient of polarity (*Eugenia jambos*);[101] development of these embryos cannot be referred to any regular type of embryogenesis because of deviations from all known types (*Ferronia limonia, Pachira oleaginea, Vincetoxicum officinale*).[123,203-205] Adventive embryos are believed to be of irregular shape and to take a lateral position with regard to the egg apparatus cells (*Citrus, Mangifera indica*).[61] Differentiation of adventive embryos does not begin simultaneously and therefore in the seed one can encounter embryos in different stages of development (*Opuntia dillenii*).[96,97]

The study of nucellar embryo development presents certain difficulties stemming from unstable position of embryocytes with regard to the embryo sac and ovule structures. It is most important to account for coincidence of the section plane with that of longitudinal axis of the embryo. To study embryogenesis median sections have been used.

Examination of embryogenesis in the genera *Sarcococca, Opuntia, Poncirus, Citrus, Zanthoxylum, Euonymus, Ochna,* and *Vincetoxicuin* indicates that first division of the embryocyte in most cases is equal; it is uncertain in direction (Plate 1, parts 1 and 2; Plate 4, part 3). Later an increasing number of cells in adventive embryos is accompanied by an increasing number of variations with respect to direction and sequence of cell divisions (Scheme I, parts 21 to 29; Scheme II, parts 14 to 16; Scheme V, parts 19 to 23; Scheme VI, part 5; Scheme 7, part 2). In many cases there is no distinct subdivision of the embryo into basal and terminal parts which are usually designated as suspensor and proper embryo parts. If these regions are singled out, the number of layers and cells is not constant. Within the same species either basal or terminal parts can develop more intensively; even in one individual they might be of different size and shape. Embryos at these stages of development are asymmetrical and are not always bipolar structures; their shape is often irregular. Despite all this, in nucellar and integumentary embryos the number of cells is gradually increasing and their shape approaches a globular one (Scheme I, parts 30 to 34; Scheme II, parts 15 to 17; Scheme V, parts 30 to 35). This shape of the embryos is reached at a different though invariably large number of cells.

Suspensor in nucellar and integumentary embryos of the species under study is frequently not pronounced morphologically when there is no clear distinction between basal and terminal parts of the embryo body. While signs of this distinction are present, the suspensor is noticeable but the number of cells and its shape are not constant and differentiation of the suspensor is possible at late stages of development (Scheme I, parts 21 to 38; Scheme II, parts 14 to 19; Scheme V, parts 19 to 37; Plate 1, part 4; Plate 3, parts 11 and 12; Plate 4, part 11).

The views of investigators on the development of the suspensor in nucellar and integumentary embryos are controversial. Some authors believe that adventive embryos are lacking suspensor and this is an important criterion to distinguish them from embryos of sexual origin (*Spatiphyllus patinii, Murrayi koenigii*).[206,207] Comparison of adventive embryo development with embryoids *in vitro* has led to the same conclusion.[208] Other scientists report that suspensor can be formed in adventive embryos. *Mamillaria tenuis* and *Toddalia asiatica* are given as examples.[209-211]

There is a third view that in species of the same genera adventive embryos can develop either with or without suspensor (*Poa pratensis*).[181,182] We think that all these conflicting views reflect the actual diversity in the development of adventive embryos.

In summary, at early developmental stages of nucellar and integumentary embryos before embryoderm formation, no regularities are observed in their divisions with respect to direction and sequence. In these cases it is impossible to be certain about the origin of cells at that or other developmental stage or about their number in embryos of a given generation. It is also unfeasible to predetermine the role and function of separate cells in forming the organs of adventive embryos.

A globular stage is of particular importance for nucellar and integumentary embryos. At this stage adventive embryos, like sexual ones, are forming embryoderm which in this case is a first indication of differentiation. Following embryoderm formation, a regular organogenesis starts in nucellar and integumentary embryos (Scheme I, parts 34 to 38; Scheme II, parts 17 to 19; Scheme V, parts 33 to 37). Divisions at the root pole become regulated; initial cells of the periblem and central cylinder are visible. Procambial cells are differentiated; tissue flattening is then observed in the terminal part of the embryo, followed by cotyledon formation. We have not observed deviations from normal organogenesis in adventive embryos. The most probable interpretation is that adventive embryos are usually located in the embryo sac cavity since the globular stage and their behavior becomes embryo sac controlled.

According to the data on *Sarcococca humilis* and *Opuntia elata*, the rate of adventive embryo development after formation of embryoderm is high; the differentiation and formation of their organs are completed during a short interval. A phase of growth of adventive embryos following that leads to their complete ripening. Such a sequence of differentiation and organogenesis usually occurs in embryogenesis of zygotic embryos of flowering plants as well.

There are practically no data on the types of development of sexually arising embryos in species characterized by nucellar and integumentary embryony. It is indicated only for some representatives that such embryos have a strict gradient of polarity and a certain sequence in divisions, but the embryogenesis type is not identified (genus *Zeuxine*).[202] In *Momordica charantia* a sexual embryo is known to develop following the *Solanad*-type and in *Beta lomatogona* — by the *Chenopodiad*-type.[119] There are also indications that suspensor is always formed in zygotic embryos. We have examined the development of zygotic embryos in *Euonymus europaea* (Scheme V, parts 10 to 18), *Opuntia ficus-indica* (Scheme II, parts 3 to 10), and *Poncirus trifoliata* (Scheme III, parts 23 to 33). The development of an embryo that does not follow classical types is called irregular. A review of works by Poddubnaja-Arnoldi[67] shows that irregular embryogenesis is described only in 2% of angiosperm families. Representatives possessing this embryogenesis type are found in different angiosperm groups, e.g., primitive taxa (families Magnoliaceae, Lauraceae, and Proteaceae), parasite plants (families Balanophoraceae, Rafflesiaceae, and Orobanchaceae), and others. It is apparent from the insignificant number of flowering plants with irregular embryogenesis that this is not the main line of development of zygote embryos in angiosperms.

The data on nucellar and integumentary embryo development show that in their development there is no close analogy with any types of zygotic embryogenesis. Differences primarily concern early developmental stages though late embryogenesis has some features in common.

The relationship between the origin and further behavior of embryos can be established by comparison of available data on adventive embryo development and existing types of embryogenesis.[212] Nucellar and integumentary embryos have been shown to arise from embryocytes located outside the embryo sac and consequently early developmental stages of these embryos occur under conditions drastically different from those of zygotic embryos. This feature is of importance because it controls the pattern of embryo development. Recent ultrastructural and biochemical investigations have revealed that the endosperm at its early developmental stages, apart from the nourishing function, is a hormone regulator for the growing embryo. Ripe endosperm does not display a strong stimulating effect.[213] In addition, it is well known that polarity of the embryo sacs and a trace of nutritive substances are in close relations. With respect to the proper embryo we should note the role of suspensor in hormone synthesis.[214-216] Thus, adventive embryos are not influenced by some regulating factors, while zygotic embryos are.

Another essential factor which controls the development of nucellar and integumentary embryos is the embryocyte itself. Unlike the zygote, this cell lacks distinct polarization of protoplast and definite position in the ovule tissue; it can be located at different angles and at varying distance from the embryo sac. For this reason adventive embryos arising by equal division of the initial cell are not bipolar structures at early stages, in contrast to zygotic embryos.

We believe that two main factors primarily affect early embryogenesis of adventive embryos: initial development outside the embryo sac and nonidentity of the embryocyte structure to that of the zygote. Thus, classification of embryogenesis should naturally incorporate the embryo origin.

Similarity in the development of adventive embryos observed in representatives of different taxa of angiosperms seems to be of importance. Those limited data now available on the development of adventive embryos still cover a number of species from various angiosperm families: *Oenothera lamarckiana* — Onagraceae; *Opuntia dillenii* — Caciaceae; *Aegle marmelos, Mamillaria tenuis, Ferronia limonia* — Rutaceae; *Bombacopsis glabra, Pashira oleaginea* — Bombacaceae; *Euphorbia dulcis* — Euphorbiaceae; *Zeuxine* — Orchidaceae; *Eugenia jambos* — Myrtaceae; *Vincetoxicum* — Asclepiadaceae; *Mangifera indica* — Anacardiaceae.[61,96-98,101,102,123,174,175,202-205,210] These results are supplemented by our covering nearly 20 species from various angiosperm families (Buxaceae, Cactaceae, Rutaceae, Celastraceae, Ochnaceae, and Asclepiadaceae), though the data on embryology are often fragmentary and do not cover all developmental stages. The contentions reported in literature are consistent with ours: an irregular pattern of early embryogenesis and a regular one of late embryogenesis.

Thus, information available on differentiation of initial cells and development of nucellar and integumentary embryos provides evidence for identical development of these structures in different taxa of flowering plants. Consequently, the development pattern of nucellar and integumentary embryos does not depend on the systematic position of a taxon characterized by this type of apomixis.

A certain similarity is also shown to occur in zygote embryo development: it is governed by five main laws of embryogenesis, irrespective of embryogenesis types.

Thus, there are two distinct lines of embryo development in angiosperms associated with different origin: (1) from the zygote and (2) from nucellus and integumentary cells (Figure 5). At present few data are available on the nature of the development of other apomictic embryos originating in the embryo sac, in particular from unfertilized egg cells, synergids, or antipodals. Therefore, discussion of their embryogenesis is impossible, though on the basis of early stages of embryo development in the embryo sac one can assume that they would be somewhat similar to zygotic embryos.

It seems reasonable to distinguish two modes of embryogenesis occurring in flowering plants: that of zygotic embryogenesis including the known types and that of nucellar and integumentary (adventive) embryogenesis as a type united for all angiosperms.

Representatives of each class can be found both in genetically close and distant taxa of angiosperms; thus, each class can exhibit an example of parallel evolutionary lines of embryo formation.

X. POLYEMBRYONY

The production of more than one embryo in a single seed is a typical feature of adventive embryony. Starting with early stages of seed development, many embryos are produced (Plate 2, parts 6 and 7; Plate 3, parts 10 to 14; Plate 4, parts 5, 11, and 12; Plate 5, part 3; Scheme VI, part 5). Polyembryony of ripening seeds is also observed (Plate 1, part 4; Plate 4, parts 13 and 14; Scheme V, parts 38 to 42; Scheme VII, part 3). The portion of seeds with several embryos in species capable of adventive embryony is often very high and can read 45 to 70% or even more. This event is exemplified in the species *Eugenia heyneana, Citrus* varieties, *Vincetoxicum officinale, Poncirus trifoliata,* etc.[100,167,217-220] Practical application of polyembryony is discussed elsewhere.[221]

Embryological investigations and analysis of ripe seeds show that the embryo number in a seed in species with this apomixis type often exceeds two and can be three to five; size and differentiation of these embryos can vary within the same seed (Plate 1, part 4; Plate 2, part 7; Plate 3, parts 11, and 12; Plate 4, parts 11 to 13). The percentage of polyembryonic seeds in species with either nucellar embryony (genera *Sarcococca, Opuntia, Citrus,* and *Poncirus*) or integumentary embryony (genera *Euonymus* and *Vincetoxicum*) is high and in most cases exceeds 50%. Instances where the monoembryonic seed formation takes place when a single embryo is always of apomictic origin are rare (e.g., *E. japonica*).[88]

In other apomixis types, when embryos can arise only from the embryo sac cells (egg cells, synergids, and antipodals), formation of more than three embryos is very unusual. Normal development of two or more embryo sacs (with embryos in each of them) can be hardly observed within a single ovule. Embryos originate from synergids and antipodals, particularly haploid ones, and stop their development before maturation. The fraction of polyembryonic seeds with these types of apomixis is small in natural conditions and rarely achieves 10%. A more frequent occurrence of polyembryony can be met in the varieties and lines in experimental breeding.

Consequently, a large portion of seeds with several embryos and a different degree of embryo differentiation in single seeds are the primary characteristics of adventive embryony. We assume that the capability of nucellar and integumentary embryony in species can be determined as the result of mature seed

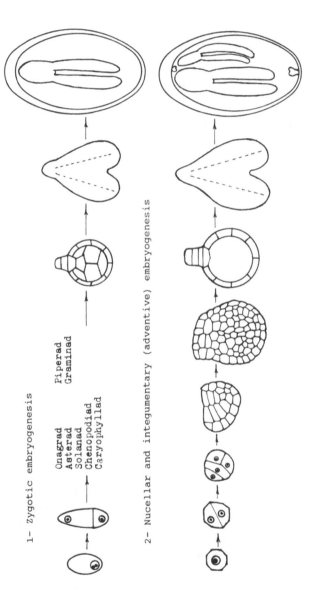

FIGURE 5. Embryogenesis in angiosperms: classification.

---

analysis, without any embryological investigation. This statement might be helpful in the search for species with adventive embryony among angiosperms, including economically important plants. It should be remembered however that one of the embryos of a polyembryonic seed might be zygotic. Possible concurrent development of embryos from the zygote and nucellus or integument cells is reported in *Spathiphyllum patinii, Ruta patavina, Murraya koenigii, and Citrus* species.[162,206,207,222] In other species degeneration of zygotic embryos and development of exclusively adventive embryos are noted in *Citrus* varieties, Mango varieties, *Pachira oleaginea, Hosta coerulea, Euphorbia dulcis,* and *Aphanamixis polystachya.*[102,105,123,139,165,223-227] There are indications that in separate species nucellar embryos are degenerating and sexual embryos are formed (*Momordica charantia.*)[124] Descriptions of embryos arising from the egg cell in the course of apomixis are given for *E. dulcis, Calostemma purpurea, Ochna serrulata, Allnus rugosa, Atraphaxis frutescens,* and *Elatostema pedunculata.*[19,133,167,170,228-230] These embryos can be formed in 8 to 10% of species capable of adventive embryony. However, in the case of nucellar and integumentary embryony, zygotic and other apomictic embryos (from synergids and antipodals) are frequently absent and only adventive embryos are developing in the polyembryonic seeds.

XI. THEORETICAL GROUNDS FOR NUCELLAR AND INTEGUMENTARY EMBRYONY

Adventive embryony is known to occur only in angiospermous plants, though its hypothetical existence can be assumed in all seed plants because adventive embryos arise from separate nucellus or integument cells of the ovule with final production of the seed. It would be important to compare peculiar features of ovule development in seed ferns, gymnosperms, and angiosperms.

It was established by Hofmeister in the middle of the last century that the nucellus of the ovule is the megasporangium or a spore sac. The main function of the integuments is to protect developing megaspores and later the seed from an unfavorable environment, though its origin is obscure. The nucellus of flowering plants might be solid (e.g., family Casuarinaceae) or thin (family Asclepiadaceae). The megasporangium (nucellus) of gymnosperms is rather uniform in structure, large in size, and consists of numerous cells. According to published data, the female gametophyte of gymnosperms are characterized by several features not encountered in angiosperms. Early differentiation of the nucellus and sporoderm formation around the female gametophyte seem to be the key features.

Information on the megasporangium structure in seed ferns and existent gymnosperms is limited. Kordyum[231] analyzed the available data and concluded that megasporangium of fossil seed plants has much in common with those of modern gymnosperms. The more complex structure of their megasporangium probably indicates a larger degree of differentiation of the nucellus and female gametophyte as compared with angiosperms.

Data on the genesis of the nucellus and the integuments in angiosperms can be summarized as follows. At the time of ovule formation, megasporocyte differentiation, and during later stages right to the formation of developed embryo sacs, the ovule cells change, but slightly. The nucellus cells (in species with crassinucellate ovules) and integument cells (with tenuinucellate ovules) are small in size, have large nuclei, dense cytoplasm, and lack supplying substances. It follows that these tissue cells remain weakly differentiated throughout the above period. A drastic change occurs in the ovule tissue after flowering and fertilization: the ovule and embryo sac start intensive growing, the nucellus and integument cells become differentiated and increase in size, degree of vacuolization is growing, and supplying substances are present. These features are clearly distinct in the cells near the embryo sac; adjacent cells degenerate. These principles can be applied to all angiosperms.

In most angiosperms initial cells become morphologically discernable from other cells shortly after flowering (during ovule cell differentiation). It is possible that their early stages of development take place when the nucellus or integumentary cells are weakly differentiated. According to existing notions weakly differentiated cells are polyfunctional. Distinctly differentiated cells can perform only one function. That means that weakly differentiated cells are totipotent and capable of taking any of the genetically coded ways of development.

If the nucellus of the ovule in angiosperms is recognized as a homologue of megasporangium, all of its cells are potentially generative structures. However, in angiosperms this mode of development is also possible for cells of the inner integument, which is revealed in particular in integumentary embryony. This can be used in support of a similar origin of the nucellus and the inner integument, though in the course of evolution there were genome changes which resulted in an ever-decreasing number of the ovule cells capable of generative functions. This is confirmed by the presence of multicellular archesporium in primitive angiosperm taxa and unicellular archesporium in evolutionary advanced plants. The majority of megasporangium cells in the latter plants function as a nutritious tissue. Solid nucellus and multicellular archesporium are adopted as ancestral types.

Two genetically coded ways of ovule cell differentiation are possible in all angiosperms. In case of amphimixis one or several subepidermal cells become generative structures (archesporial cells). In certain types of apomixis the sporadically located cells can function as generative structures. This process takes place at different stages of ovule ontogenesis: emergence of the above cells at early stages leads to apospory although their emergence at later stages gives rise to adventive embryony.

In the case of apospory and adventive embryony, more cells of the ovule can be generative structures as compared with amphimixis. Determination of embryocytes requires lasting differentiation of the nucellus and integument tissue. The possible consequence is frequent occurrence of adventive embryony among trees, shrubs, and perennials characterized by slow development of generative structures. It seems likely that in amphimictic angiosperms the differentiation capacity of initial cells of adventive embryos and aposporic embryo sacs can be also manifested if the suppressing factor is no longer operative. Early differentiation of the nucellus in gymnosperms appears to be the main obstacle to the above apomixis types.

Against this background the differentiation of initial cells of aposporic embryo sacs and embryocytes does not seem unusual. In gymnosperms a sporoderm surrounding the female gametophyte is probably another obstacle for adventive embryo development. The sporoderm securely isolated the gametophyte due to its thickness, toughness, and composition, as does the pollen grain envelope. In contrast to the sporoderm, the embryo sac cell wall of angiosperms is similar in its structure to ordinary cell walls of plant tissue (except for the absence of plasmodesmata). This envelope does not prevent adventive embryos from penetrating the embryo sac.

Thus, comparison of the megasporangium and female gametophyte in gymnosperms with the ovule and embryo sac in angiosperms reveals their structural and functional differences which are responsible for the occurrence of adventive embryony only in angiosperms.

XII. ADVENTIVE EMBRYOS AND EMBRYOIDS

Recent wide application of a tissue culture technique has allowed the obtaining of embryoids — embryo-like structures (somatic embryos). We will briefly compare the behavior of adventive embryos and embryoids formed *in vitro*. In both cases a new organism arises not from the gamete and in the absence of fertilization, though in the first instance this process takes place in natural conditions and is an indication of the species, while in the second instance it is of artificial origin.

The main features of embryoids have been discussed in a number of works.[80,232-241] However, to obtain embryoids in experiment is still a difficult task because there is no comprehensive understanding of the conditions responsible for embryoidogenesis; it particularly concerns early developmental stages. Nobody can be confident which initial cell would later give rise to the embryoid, though presumably they are dense cytoplasmic cells of suspension or callus. These cells in callus can be either epidermal or subepidermal. Ultrastructure of these cells is briefly discussed in a few works.[235,240,241] Supposed embryoidogenic cells are believed to be meristematic, have a large nucleus, and a cytoplasm with numerous organelles. There is no unanimous opinion regarding plasmodesmata: either they are present or not functional, if present at all. First division of the embryoid initial cell is thought to be unequal; one of two daughter cells is smaller, undergoes an intensive division, and produces the proper embryo, although the other is divided more slowly and forms a suspensor. However, successive stages of embryoid development have not yet been investigated. Since initial cells of embryoids are not identified with certainty, it is perhaps premature to make a judgment on the course of the first division of this cell. Our analysis of published data shows that a distinct polarity is typically absent in the supposed initial cell of the embryoid. If this is the case it is reasonable to assume that the initial cells of nucellar and integumentary embryos have features in common with those of embryoids: the presence of cell organelles abundant in cytoplasm and the absence of the clearly distinct polarity. However, in this case an unequal division of the cell cannot occur. It is also of note that embryoids are reported to originate both from separate cells and from groups of cells.[181,239] Our study of adventive embryos shows only one way of formation in nature — from a separate initial cell. This implies that adventive embryos and embryoids are not always formed in a similar fashion.

Many investigators have indicated there are no regularities at early stages of embryoid development. Nominal and successive cell divisions are typically absent, as noted earlier in adventive embryos. The absence of distinct polarity in initial cells and developing adventive embryos, combined with development outside the gametophyte, is probably responsible for such a specific and similar pattern of these embryos and embryoids at early development. Likewise, conclusions have been made earlier by Haccius[242] and other scientists. As shown above, adventive embryo development after embryoderm differentiation is governed by general regularities of differentiation and organogenesis which take place in embryos of sexual origin. We have not recorded any abnormalities and teratological cases in adventive embryos of these stages. Differentiation and organogenesis also occur during late development of embryoids, but abnormal deviations are quite common. Some authors even believe that embryoids *in vitro* lack differentiation despite the external likelihood with sexual embryos.[240] This statement does not seem quite accurate and evidence for certain signs of differentiation and organogenesis in embryoids is provided in work by Halperin[232] and other investigators.

Therefore, comparison of embryoids with adventive embryos reveals some common features in their development and differentiation of their initial cells.

Recently, some positive results have been achieved in experiments on tissue culture of ovules. In particular the nucellus of some citrus plants has been successfully used *in vitro* for microclone multiplication and virus-free lines of nucellar plants.[235,244] Experimental growing of the ovules in culture under controlled conditions has been started in the area of forest biotechnology to obtain pure variety lines characterized by disease resistance and great hardiness.[245] Attempts to obtain embryoids from the nucellus of some Cucurbitaceae and several representatives of the genus *Gossypium* are yet unsuccessful.[238] Experiments with *Citrus* seem to reveal that in plants characterized by adventive embryony the nucellus and integument cells *in vitro* are capable of undergoing embryoidogenesis more easily than other plants. At present the ovules of species with adventive embryony are used in this field on a limited scale but probably the majority of species in our list (page 63) will behave like citrus plants and will be promising objects in tissue culture.

Application of apomixis for plant hybridization is of lasting interest among investigators. Karpechenko[246] was among the first to have shown its possible application to maintain heterosis in cultivated plants. With this end in view adventive embryony is most suitable due to the heredity of nucellar and integumentary embryos on the mother line. We believe that nucellar and integumentary embryony can be induced much more easily than other types of apomixis. The main prerequisite is totipotentiality of the ovule cells and the course of transfer to apomixis in this case includes only one stage — differentiation of the initial cells of adventive embryos. Experimental studies in the area of adventive embryony are promising and desirable because it occurs in a number of economic plants. Seedlings originated from nucellar and integumentary embryos are shown to have several important advantages over sexual embryos: they possess a great hardiness, can resist viruses, and do not inherit the disease from a mother organism.[238]

Chapter 4

OCCURRENCE OF NUCELLAR AND INTEGUMENTARY EMBRYONY AND ITS EVOLUTIONARY SIGNIFICANCE

I. OCCURRENCE OF NUCELLAR AND INTEGUMENTARY EMBRYONY IN THE SYSTEM OF FLOWERING PLANTS

To assess the significance of a phenomenon one should know not only its structural background but also its frequency of occurrence. Our list of species with adventive embryony has been compiled from both published and original data. The sequence of families and enumeration of genera and species within families follow alphabetical order. The taxa are in accordance with the system of Takhtajan.[247]

Adventive embryony is found in 250 species of 121 genera belonging to 57 families united in 35 orders of flowering plants. However, this number of species probably does not cover all species capable of this type of apomixis since investigations involve only a small quantity of plants. The data obtained can be considered a random sampling which provides some information on the properties of a general set, i.e., flowering plants as a whole. Data on occurrence of apomixis in angiosperms were given by Khokhlov et al.[57] The authors have shown that apomixis (without accounting for adventive embryony) is observed in approximately 700 species from 377 genera belonging to 97 families. A comparison of our data with that of Khokhlov shows that adventive embryony occurs in 59% of families covering 66% of genera and 34% of species. Therefore, adventive embryony is one of the most widely spread types of apomixis in angiosperms.[248]

Analysis of distribution of adventive embryony in the orders according to Takhtajan's scheme[247] has shown that it is found sporadically throughout the phylogenetic tree, without any restriction (Figure 6). Adventive embryony is found in 37% of angiosperm orders. This phenomenon was traced in 206 species of 91 genera referred to 45 families of dicots and in 41 species from 30 genera of 12 families of monocots.

The list of species with adventive embryony provides additional information on this apomixis type. First, nucellar and integumentary embryony is mostly found in certain life forms: trees and shrubs — 73%, perennial grasses — 22%, and biannuals — 5%. It was not known in annuals. Also, in our list of species plants of a temperate zone are rather rare, e.g., *Poa pratensis, Colchicum autumnale, Potentilla geoides,* genera *Malus* and *Beta;* for the most part they belong to tropical and subtropical zones — *Mangifera, Citrus, Opuntia, Pachira, Sarcococca, Vincetoxicum, Combretum, Garcinia, Cassia,*

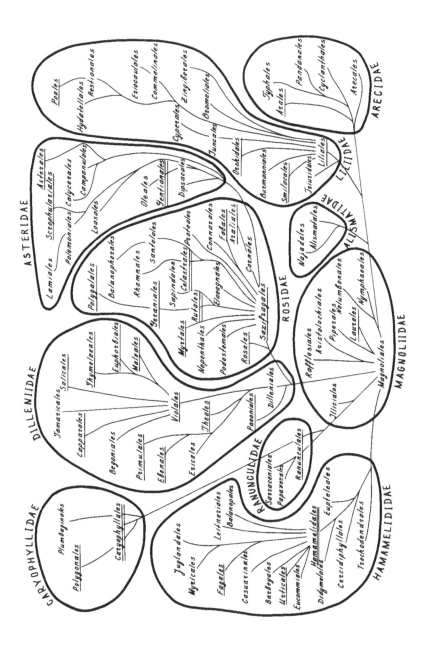

FIGURE 6. Putative relationships among the classes, subclasses, and orders of flowering plants. – Orders with representatives capable of adventive embryony; = Orders with representatives of capable adventive embryony, as investigated by Naumova. (From Takhtajan, A., BOREA. 46(3) 226. 1980. With permission.)

Eugenia, Fortunella, and *Sizygium.* Distribution areas of others partially extended from a subtropic to a temperate zone (*Euonymus*). Cosmopolites with adventive embryony are extremely rare.

Data on somatic chromosome numbers are of substantial interest. The opinion exists that species with adventive embryony are often diploids, and this type of apomixis is more easily achieved at the diploid level, while other types of apomixis prevail in polyploids.[7,12,17,20-22,24,25,51,53] Our analysis of chromosome data has shown that a majority of species with adventive embryony possess high numbers of somatic chromosomes probably being polyploids; euploids and aneuploids also occur, while diploids are rare. The occurrence of polyploids in plants with adventive embryony enables one to suppose that hybridization might be one contributing factor of their emergence. Our conclusions are consistent with data on an Australian species of *Cassia*.[249] The author has found that nucellar embryony is rare in diploid species and frequent in polyploid ones.

II. NUCELLAR AND INTEGUMENTARY EMBRYONY WITHIN FAMILIES, GENERA, AND SPECIES OF FLOWERING PLANTS

Nucellar and integumentary embryony most frequently are met in the families Rutaceae (10 genera, 35 species), Orchidaceae (10 genera, 13 species), Malpighiaceae (7 genera, 8 species), Poaceae (7 genera, 8 species), Rosaceae (4 genera, 11 species), Myrtaceae (3 genera, 13 species), Cactaceae (3 genera, 9 species), and Celastraceae (2 genera, 19 species). For some families mentioned above there are additional data which are useful.

Analysis of the generic and specific compositions of the family Buxaceae, its distribution, paleobotany, and evolution with respect to anatomical-morphological characteristics reveals its ancient origin.[250] Our data on reproductive structure development support this view and testify that the genus *Sarcococca*, possessing nucellar embryony, is probably one of the most primitive in the family. The abnormalities in the development of reproductive structure are usual, while the ability for seed formation is maintained only by apomixis.

In Cactaceae nucellar embryony is known in the genera *Opuntia* and *Mammillaria*. In all five species of *Opuntia* studied so far only this form of apomixis is noted in *O. vulgaris, O. ficus-indica, O. leucantha, O. aurantiaca,* and *O. dillenii*.[35,36,58,94-97,251] In *O. ficus-indica* the occurrence of a zygotic embryo along with nucellar ones is assumed. Our observations show a rare occurrence of amphimixis in *O. elata* and nucellar embryony is predominant. The presence of amphimixis in *O. ficus-indica* and *O. elata* demonstrates sexual and asexual modes of seed formation. Regrettably, only specimens of *Opuntia* from greenhouses have been studied; more diverse forms are found in the wild. It seems probable that *Opuntia* has amphimixis, nucellar embryony, and vegetative multiplication. Coexistence of three modes of reproduction contributes to the entity of their extensive distribution area.

In Rutaceae nucellar embryony occurs in many species from the genera *Aegle, Citrus, Clausena, Esenbeskia, Feronia, Fortunella, Myrraya, Poncirus, Ptelea, Triphasia,* and *Zanthoxylum.* In all genera, except *Zanthoxylum,* embryos can be formed from zygotes or nucellus cells. Nucellar embryony in *Citrus* is significant for seed formation. Seeds with nucellar embryo are more numerous than those of zygotic origin. Therefore, seed production is achieved by two mutually supplementary processes: amphimixis and nucellar embryony. Nucellar embryony is a single mode of seed formation in all studied *Zanthoxylum: Z. bungei, Z. alatum, Z. planispinum, Z. americanum,* and *Z. simulans.*[43,103,104,194,252] According to Fish and Waterman,[253] this genus contains nearly 15 species. Five (30%) were studied embryologically; in all five species seed production was maintained due to nucellar embryony.

Formation of fertile pollen grains in *Z. americanum* and *Z. schinifolium,* growing even outside their geographical distribution, does not allow referring of *Zanthoxylum* to obligate apomictics.[194] Thus, in *Zanthoxylum* and in other Rutaceae members the possibility of amphimixis occurrence cannot be neglected. Our data on megaspore tetrads collapse and the formation of aposporic embryo sac in *Z. americanum* fits well into Mauritzon's observation, assuming apospory in *Z. alatum* and *Z. bungeanum.*[252] The formation of aposporic embryo sacs in *Zanthoxylum* seems more sound. In the case of aposporic embryo sacs and fertile pollen grains, one can suppose formation of triploids by fertilization or diploids without fertilization. According to taxonomical data *Zanthoxhylum* is considered the most primitive in Rutaceae and moreover as a direct line, connecting Rutaceae with the order Ranales.[253,254] It seems probable that the occurrence of apospory and nucellar embryony in *Zanthoxylum* reserves the ability to form seeds in the case of a lack of or a rare occurrence of amphimixis. This factor is very important for arborescent life forms where vegetative multiplication is restricted.

The occurrence of adventive embryony and peculiarities of embryology in very few families (Buxaceae, Cactaceae, Rutaceae) prevent one from making some general statements. It seems probable that adventive embryony occurs in genera of various specialization levels. In advanced genera deviations in reproductive structure development are less pronounced while seed reproduction is achieved by amphimixis and adventive embryony and in some cases other types of apomixis occur. The most primitive genera embryology — *Sarcococca* and *Zanthoxylum* — deviate in many respects. While seed reproduction is achieved by adventive embryony and sometimes by additional types of apomixis, the sexual process is not found.

The occurrence of adventive embryony within genera of important phylogenetic data is needed for this kind of study. Distribution of integumentary embryony was analyzed in the genus *Eonymus.*[255] To trace polyembryony and integumentary embryony (other types of apomixis in this genera are lacking)

TABLE 2
Number of Embryos in Mature Seeds of Native Species of the Genus *Euonymus*

Species	Section[a]	Number of embryos in mature seed
1. *Euonymus europaea*	*Euonymus*	One
2. *E. czernjaevii*	*Euonymus*	One
3. *E. velutina*	*Euonymus*	One
4. *E. maackii*	*Euonymus*	One
5. *E. sieboldiana*	*Euonymus*	One
6. *E. pauciflora*	*Pseudovyenomus*	One
7. *E. verrucosa*	*Pseudovyenomus*	One
8. *E. semenovii*	*Pseudovyenomus*	One
9. *E. nana*	*Pseudovyenomus*	One
10. *E. koopmani*	*Pseudovyenomus*	One, two, or more
11. *E. alata*	*Melanocarya*	One, two or more
12. *E. sacrosancta*	*Melanocarya*	One
13. *E. latifolia*	*Kalonymus*	One, two, or more
14. *E. maximovicziana*	*Kalonymus*	One, two, or more
15. *B. planipes*	*Kalonymus*	One, two, or more
16. *E. macroptera*	*Kalonymus*	One, two, or more
17. *E. leiophloea*	*Kalonymus*	One, two, or more
18. *E. sachalinensis*	*Kalonymus*	One, two, or more
19. *E. miniata*	*Kalonymus*	One, two, or more

[a] Sectional attribution of species followed by Leonova.[256]

mature seeds of all 19 species of *Euonymus* native to the USSR were investigated. Seeds of some introduced species were studied. Tables cited below are compiled from published and original data. In 8 of 19 species native to the USSR seeds with several embryos were found. Therefore, integumentary polyembryony takes place (Table 2). As shown in Table 3, for 9 of 16 *Euonymus* species polyembryony is typical. Therefore, in 17 of 35 species the tendency toward integumentary embryony is observed. However, earlier works have shown that integumentary embryony can be met in species with one embryo, as in *E. japonica*[88] or in *E. europaea*.[70] According to *Leonova*, [256] species in question belong in terms of systematics to two subgenera, *Euonymus* and *Kalonymus,* covering nine sections (Figure 7). On the basis of study of mature seeds and embryology, polyembryony was reported in species from the sections *Pseudovyenomus, Melanocarya, Echinococcus, Vyenomus, Kalonumus, Euonymus*, and *Ilicifolia.*

So, integumentary embryony is found in seven of nine sections of the genus. This occurs in sections with native species of genera *Euonymus*. Only in the sections *Hymphrya* and *Hesperidionymus* is no integumentary embryony found; it is probably due to the limited scope of studies. The scheme of Leonova[256] allows analysis of the integumentary embryony in a phylogenetic respect. This

TABLE 3
Number of Embryos in Mature Seeds of the Genus *Euonymus*

Species	Section	Number of embryos in mature seed	Ref.
1. *Euonymus bungeana*	*Euonymus*	One	255, 257
2. *E. hamiltoniana*	*Euonymus*	One	255, 257
3. *E. atropurpurea*	*Humphrya*	One	258, 259
4. *E. japonica*	*Ilicifolia*	One	88, 255, 257
5. *E. occidentale*	*Hesperidionymus*	One	258, 259
6. *E. ovata*	*Hesperidionymus*	One	258, 259
7. *E. alata var. apretus*	*Hesperidionymus*	One, two or more	258, 259
8. *E. alata f. angustifolia*	*Hesperidionymus*	One, two or more	255, 257
9. *E. verrucosoides*	*Hesperidionymus*	One, two or more	258, 259
10. *E. americana*	*Echinococcus*	One, two or more	260
11. *E. dielsiana*	*Vyenomus*	One, two or more	258, 259
12. *E. vagans*	*Vyenomus*	One, two or more	258, 259
13. *E. fimbriata*	*Kalonymus*	One, two or more	260
14. *E. yedoensis*	*Kalonymus*	One	255, 257
15. *E. latifolia*	*Kalonymus*	One, two or more	29, 129,261
16. *E. oxyphylla*	*Kalonymus*	One, two or more	258, 259

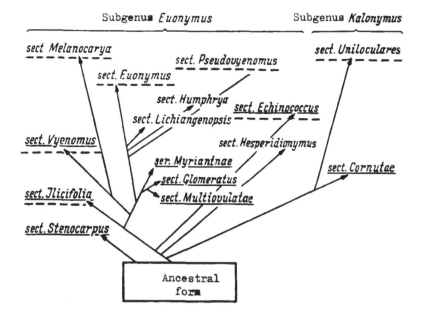

FIGURE 7.   Phylogenetic relationships in the genus *Euonymus*. – Sections with evergreen species. ---- Sections with representatives capable of integumentary embryony. (From Leonova, T., *Euonymus Native to the U.S.S.R. and Neighbouring Countries*, 1974, 132. With permission.)

form of apomixis occurs sporadically throughout the taxa. Representatives of most sections at the scheme base are evergreen; ancestral types were supposed to be evergreen. The tendency toward integumentary polyembryony was registered for evergreen species - *E. japonica* (section *Ilicifolia*), *E. dielsiana* (section *Vyenomus*), and *E. americana* (section *Echinococcus*). The most ancient and primitive is section *Ilicifolia*. In *Euonymus* species integumentary embryony was found in sections of subgenus *Euonymus* and in subgenus *Kalonymus;* the latter is considered younger.[256] One can suppose that integumentary embryony can be a typical feature of genus *Euonymus* as a whole. It can originate early in ancestral form and it was fixed in generations genetically.

As the tertiary representatives of Celastraceae were widely distributed throughout the Earth, distributional area and size of family diminished later on. Nowadays the family is restricted to tropical and subtropical zones.[262] Only 2 of 58 genera (*Euonymus* and *Celastrus*) are in north temperate and subtropical zones.[256,259] These genera are capable of growing, of disseminating in the zone of temperate climate, that is, under conditions unfavorable to other members of the family. It seems probable that the ability of integumentary embryony characterizing *Euonymus* and *Celastrus* to prevent diminishing of distributional area in comparison with other representatives of the family Celastraceae lacking apomixis.

Such analysis is not feasible within other groups of flowering plants due to the lack of phylogenetic studies. However, according to our list adventive embryony is frequent in many investigated species of separate genera. One can mention, for instance, *Sarcococca* (4 species), *Opuntia* (7 species), *Citrus* (18 species), *Zanthoxylum* (6 species), *Shorea* (6 species), *Eugenia* (11 species), *Cassia* (8 species), *Ardizia* (6 species), *Vincetoxicum* (5 species), and *Ochna* (3 species). Therefore, the conclusion that adventive embryony can be a trait of the genus can be true for other genera of flowering plants. In this case, the majority of specimens possesses a high percent of polyembryonal seeds. However, adventive embryony is sometimes met only in single representatives of the genus, if present at all (genera *Ornithogalum, Paeonia, Calycanthus, Gentiana*).

The frequency of occurrence of adventive embryony within species for which this phenomenon is typical has not been studied. There are only rare publications on population aspects of adventive embryony. In *Spiranthes cernua*, North American orchid, three races were found according to the mode of reproduction base: apomictic, amphimictic, and intermediate.[115, 263] Sexual and apomictic races were registered in *Nigritella nigra*. Scandinavian races were apomictic with frequent adventive embryony. Races from the Alps were normally sexual ones.[21] Studies on the population level of genera *Rubus, Potentilla, Alchemilla,* and *Sanquisorba* show various ratios of sexual and apomictic forms within the population.[264] Using our data on *Opuntia elata* and *O. ficus-indica* and materials on adventive embryony of a population level, one can suppose within species capable of adventive embryony the existence of amphimictic, apomictic, and intermediate races. Embryological studies have shown that intermediate races with amphimixis and adventive embryony are most frequent in nature.

III. EVOLUTIONARY SIGNIFICANCE OF NUCELLAR AND INTEGUMENTARY EMBRYONY

The evolutionary significance of adventive embryony has not been discussed. Evolutionary assessment of apomixis is controversial.[7,9,10,20-22,51,265,266] There is a tendency to consider amphimixis and apomixis as a single system of reproduction modes.[267,268] The opinion exists that a combination of apomictic and sexual individuals within a population represents a very flexible adaptive system, which allows one to take advantage of the sexual process and apomixis simultaneously,[268] although the above observations are not applied to adventive embryony. Our embryological studies and published data on species capable of adventive embryony have shown the occurrence of this phenomenon in the phylogenetic system of flowering plants, as well as within families, genera, and species, contributing to our understanding of the evolutionary significance of this apomixis type. Embryological data confirm that, in the case of adventive embryony, as in other types of apomixis, both amphimixis and apomixis mostly participate in seed reproduction. Amphimixis brings about genetic diversity in progeny, while adventive embryony simultaneously with reservation of the mother genome fixes variation and stabilizes hybrids. Adventive embryony, as in *Euonymus macroptera* and *E. planipes*, maintains individual capacity for seed formation. In these species nearly 46% of seed is formed due to adventive embryony. In a few species of the genera *Sarcococca, Zanthoxylum, Zeuxina,* and *Euphorbia,* seed production is achieved by adventive embryony exclusively. The occurrence of adventive embryony is reflected in geographical distribution of taxa. Some of them which retain the capacity for amphimixis and adventive embryony have extensive and continuous distributions (genera *Euonymus, Opuntia*)., while the taxa with seed reproduction due to adventive embryony only have more restricted areas (genus *Sarcococca*).

Therefore, adventive embryony is of great evolutionary significance since it stabilizes the hybrid genotype and plays a significant and sometimes even primary role in maintaining seed productivity.

Chapter 5

APOMIXIS AND AMPHIMIXIS IN SEED
PRODUCTION OF FLOWERING PLANTS:
CLASSIFICATION

The previous chapters discussed structural and functional grounds for the occurrence of nucellar and integumentary embryony and the controversial views in the literature on the subject.

The subdivision of agamospermy into gametophytic apomixis and adventive embryony, where the latter is likened to vegetative multiplication, has recently become widespread.[12,13,51,53]

These works do not account for structural and functional regularities of seed formation in the case of apomixis, amphimixis, and vegetative multiplication. This suggests common mechanisms underlying these processes. In light of the above information (Chapter 3), the following conclusions can be drawn.

First, the nucellus and integument cells are not truly somatic, similar to cells of vegetative organs. The conception of the nucellus as megasporangium leads to an assumption that cells of the nucellus and integument (as its derivative) are generative structures in their nature. This generative capability is realized in aposporic embryo sac formation (= gametophytic apomixis) and in nucellar and integumentary embryo development (= adventive embryony). Second, the development of nucellar and integumentary embryos is fundamentally different from the bud formation that underlies vegetative multiplication. Both zygotic and adventive embryos undergo all stages of organogenesis. Sexual and adventive embryos are morphologically indiscernible in the mature seed. Some differences during early embryogenesis can be attributed to their origin and varying developmental conditions. The presence of the endosperm is indispensable for normal embryo development, irrespective of their nature (excluding some specialized taxa). Third, the data from ultrastructural investigations show that differentiation of all embryonic structures has a common pattern, whether they are formed in the course of meiosis and amphimixis or by abnormal meiosis, mitosis, and without fertilization. Our view is that the mechanisms responsible for generative structure differentiation during amphimixis and apomixis are in fact universal.

A survey of data obtained leads to the conclusion that during normal formation of generative structures regular processes take place at the above stages, characterized as follows: (1) changes in the cell wall structure manifested in the loss of plasmodesmata during megasporogenesis and megagametogenesis, and subsequent restoration of the cell wall structure at the zygote and zygote embryo; (2) changes in ribosome population number; (3) changes in morphology of plastids, mitochondria, endoplasmic reticulum, etc.; and (4) reorganization in nuclei ploidy. Reorganization of protoplast structure during alternations of development phases eventually results in changing inter-relations between the cells and changing the metabolism with adventive embryony it is an interesting point, whether the protoplast reorganization of generative cells can occur in the absence of alternations of generations (sporo-phyte-gametophyte) and nucleus ploidy.

Ultrastructural studies on generative cell protoplasts of apomictic plants are few in number and deal with adventive embryony, and in part with apospory.[144-148,269] The reorganization of the initial cell protoplast in the case of adventive embryony, as was shown, is comparable in general with that of generative cells in the case of amphimixis (plasmodesmata loss, morphological changes in cell organelles, etc.), despite the occurrence of nucleus ploidy changes and alternation of generations in the latter instance and its nonoccurrence in the first instance.

It is important to clarify also whether protoplast transformations can take place during the formation of a new individual without fertilization. Our data on synergid differentiation capable of producing an embryo without fertiliza-tion show that the same transformations in the cell wall structure are observed, likewise during normal zygotogenesis and embryogenesis.[62,270]

Relying on the limited data currently available, we are inclined to assume that the reorganization of generative cell protoplasts and their isolation occurs in all steps of developmental stage alternations: in sporogenesis, zygotogenesis, and, probably, it also takes place in the cases of apospory, apogamety, and parthenogenesis, and it is observed in the absence of these alternations — in the case of adventive embryony. The protoplast reorganization seems also to be independent of meiosis, fertilization, and consequently of changes in nuclei ploidy. It is evident that genetic programs of nuclei ploidy changes and proto-plast reorganization are guided by different genes.

In summary, we believe that adventive embryony and gametophytic apo-mixis are not fundamentally different types of agamospermy.[366] The view that adventive embryony is closely related to vegetative multiplication is hardly acceptable. The absence of the above alternations of the generations in the case of adventive embryony is not the character that prevents its inclusion in a general system of apomixis along with its other types and forms.

The present classification is based on two principles: (1) the nucellus is treated as megasporangium whose cells are potentially generative and have specific features in differentiation during the transition to development of embryo sacs (diploid or haploid) and adventive embryos; (2) embryo sac cells are regarded as potential gametes which are capable of giving rise to embryos without fertilization. Our classification, like that suggested earlier by Gustafsson,[20] subdivides apomixis into diplospory, apospory, and adventive embryony. It also has some elements encountered in classifications of other investigators;[11,18,26,27] this concerns in particular separate consideration of sporogenesis phases and mode of embryo formation. We use the binary termi-nology proposed earlier by Modilewsky[4] and Khokhlov[23] to define the types and forms of apomixis. Specific features of deviation during sporogenesis resulting in an unreduced number of chromosomes during meiosis of megasporocytes are not indicated. Vegetative multiplication is not included in apomixis.[271]

The present classification allows demonstration of the developmental stage that leads to apomixis (Figure 8). Stages of seed development during amphimixis are presented in the left part of the scheme and types and forms of apomixis are in the right part. It is proposed to single out the initial cell stage as a necessary step in generative structure development in the case of apospory and adventive embryony. This reveals the different natures of apomixis types: some are the result of circumventing meiosis right to mitosis of megasporocytes (= diplospory), although others occur exclusively on the basis of mitosis of the ovule cells (= apospory and adventive embryony). Moreover, the notion "gametophytic apomixis", which does not reflect in our opinion the major regularities of apomictic development, can be dismissed. It seems encouraging to draw a distinction between the type and form of apomixis. The type of apomixis should be defined by division type and the nature of initial cell of the ovule entering this division. Megasporocyte, initial cell of aposporic embryo sac, and initial cell of adventive embryo (embryocyte) can be such cells. The form of apomixis is determined by the structure of the embryo sac which produces an embryo, i.e., egg cell, synergids, or antipodals. The classification discerns three types of apomixis: diplospory, apospory, and nucellar and integumentary (adventive) embryony, and also three forms of apomixis: parthenogenesis, apogamety, and gemigamy.

Diplospory (= apogamy, generative apospory, semiapospory, aneuspory, pseudoapospory) is the type of apomixis whereby the embryo sac is formed from a megasporocyte as the result of circumventing meiotic divisions (semigeterotypic, pseudohomotypic, and apohomotypic) right to the mitosis. These divisions lead to an unreduced chromosome number and possible genetic nonhomogeneity of developed embryo sacs and eventually embryos. These issues are examined in a number of works.[11,18,26,27] Embryo sacs which arise due to diplospory develop according to *Taraxacum, Ixeris,* and *Antennaria* types. Peculiar features of formation and development of these embryo sacs are discussed elsewhere[11,27,53,272] together with a list of species characterized by this type of apomixis, e.g., *Antennaria alpina, Poa alpina, P. nevrosa, Nardus stricta, Agropyron scabrum, Ixeris dentata, Allium nutans, Parthenium, Taraxacum, Chondrilla,* etc.

Apospory (= somatic apospory, somatic evapospory, evapospory, etc.) is the type of apomixis whereby the embryo sac is formed from the initial cell, a derivative cell of the nucellus, as the result of mitotic division. Cells of the embryo sac are characterized by the diploid chromosome number and heredity on a mother line. Aposporic embryo sacs develop following *Hieracium* and *Panicum* types.[11,53,340] This type of apomixis is frequently observed in representatives of the order Poales: *Poa pratensis, Hierochloe, Panicum, Sorghum,*

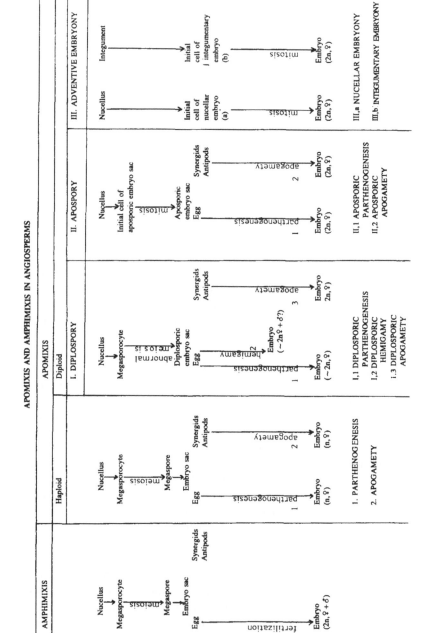

FIGURE 8. Apomixis and amphimixis in angiosperms: classification.

etc. and in some other angiosperms from the genera *Malus, Beta, Ranunculus,* etc). Our observations on the genus *Zanthoxylum* indicate that initial cells of apomictic embryo sacs at the light microscope level have much in common with those of adventive embryos. These cells can be located in different parts of the ovule and take any position except subepidermal. Initial cells of aposporic embryo sacs become discernable during sporogenesis and the development of normal embryo sacs or during their degeneration. There are several recent works concerning apospory.[272,273]

Adventive embryony (= nucellar and integunentary embryony, apogametophytic sporophyty) is the type of apomixis whereby the embryo is formed from the nucellus or integument cells which become embryocytes. As noted above, initial cells of adventive embryos are obviously similar to those of aposporic embryo sacs. Basic embryological data on this apomixis type can be found in the present work.

Parthenogenesis (= ovogen apogamy, apozygoty, unreduced parthenogenesis, reduced generative parthenogenesis, unreduced somatic parthenogenesis, pseudogamy, etc.) is the form of apomixis when an embryo arises from a haploid or diploid unfertilized egg cell. Haploid parthenogenesis can be encountered in normally developing embryo sacs with a reduced chromosome set as a result of meiosis. This form of apomixis is of rare occurrence and is found in corn and some other economic plants. Diploid parthenogenesis can occur in embryo sacs originating as a result of diplospory or apospory. It is widely spread in nature and is primarily observed in species with the above types of embryo sacs. To define diploid parthenogenesis we propose to use the following binary terms: diplosporic parthenogenesis — when embryos develop from egg cells of embryo sacs originated due to diplospory — and aposporic parthenogenesis — when embryos develop from egg cells of aposporic embryo sacs.

Apogamety (= apogamy, somatic parthenogenesis, somatic apogamy, etc.) is the form of apomixis when an embryo arises from synergid or antipodal without fertilization. Haploid apogamety is of rare occurrence and can be observed in embryo sacs with a reduced chromosome set. Diploid apogamety takes place in diplosporic or aposporic embryo sacs. A more accurate term for this apomixis form would be diplosporic or aposporic apogamety.

Hemigamy (=semigamy) is the form of apomixis when an embryo develops from an egg cell with a sperm penetrated but without fusion of their nuclei. During early embryo development the male and female nuclei start to divide independently of each other. Their further embryological development is still unclear. Hemigamy occurs in embryo sacs either with reduced or unreduced chromosome set which are brought about by diplospory (diplosporic hemigamy) and it is observed in a few species of angiosperms.[11,274,275]

The present classification, in contrast to many earlier classifications, does not subdivide adventive embryony into autonomous and induced types. Analysis of published data and our observations have shown that initial cell

formation and early development of adventive embryos are not directly dependent on pollination, fertilization, the presence of sexual embryos, or the presence of pollen tubes in the ovary or in the embryo sac. Thus, such a division
would be unjustifiable and confusing. It also seems unreasonable to divide
other apomixis forms into autonomous and induced because there is no con-
vincing experimental evidence for the dependence of apomixis on pollination.
This subdivision was first advocated by Ernst[16] as early as 1918 and became
commonly accepted. However, we hold the view that this controversial division introduces additional difficulties in apomixis studies.

Our classification does not include some apomixis forms which do not occur
in nature and are obtained experimentally, e.g. androgenesis and pseudogamy.

Androgenesis (reduced and unreduced male parthenogenesis) is the form of
apomixis whereby an embryo develops from a diploid or haploid egg cell; in
this case a sperm nucleus remains functional, while a nucleus of the egg cell
itself collapses.

Pseudogamy is the form of apomixis whereby an embryo develops from a
diploid or haploid egg cell with a sperm penetrated but fertilization does not
occur and the egg cell nucleus remains functional.

Thus, the classification suggested demonstrates the basic modes of embryo
sac and embryo formation in angiosperms. Studies in the field of adventive
embryony which are reflected in our scheme show that unification of amphimictic and apomictic modes of seed production among angiosperms into a
single system is not merely formal. They are similar not only in the final
product of seed formation but also in differentiation processes and generative
structure development. Adventive embryony is to be recognized as one of the
major types of apomixis along with diplospory and apospory.

Chapter 6

A LIST OF SPECIES WITH NUCELLAR AND INTEGUMENTARY EMBRYONY

N — nucellar embryony, I — integumentary embryony, P — parthenogenesis, Ag — apogamety, and Ap — apomictic embryo sacs (diplosporic or aposporic). The sequence of references follows chronological order.

Taxa 1	Types and forms of apomixis 2	Ref. 3
Dicotyledones		
Anacardiaceae — Rutales		
Mangifera indica	N	29, 42, 226, 227, 170, 171, 276
M. odorata	N	277, 278, 127, 142, 279
Apiaceae — Araliales		
Ammi majus	I	280
Asclepiadaceae — Gentianales		
Vincetoxicum cretaceum	I	281, 167, 205, 282
V. officinale (Cynanchum *vincetoxicum)*	I	
V. stepposum	I	
Cynanchum medium	I	
C. nigrum	I	
Asteraceae — Asterales		
Carthamus tinctorius	I	283–285
Hieracium alpinum	I	
Melampodium divaricatum	I?, Ag	
Betulaceae — Fagales		
Alnus incana	N, Ag, P	169, 170, 286
A. rugosa	N, Ag, P	
Betula oycoviensis	N, Ap, P	
B. pendula	N, Ap, P	
Duschekia kamschatica	N, AP, P	
Bombacaceae — Malvales		
Bombacopsis glabra	N	287, 123, 174
Bombax kimuensae	N	
B. oleagineum	N	
Pachira oleagina	N	
Burseraceae — Rutales		
Garuga pinnata	N	288
Buxaceae — Hamamelidales		
Sarcococca hookeriana	N	289, 89–91, 252, 290, 73, 74, 291
S. humilis	N	
S. pruniformis	N	

Taxa 1	Types and forms of apomixis 2	Ref. 3
Dicotyledones (continued)		
S. ruscifolia	N	
S. zeylanica	N	
Cactaceae — Caryophyllales		
Mammillaria tenuis	N	260, 35–37, 58, 96, 97, 94, 209
Opuntia elata		210, 251, 95, 193, 75
O. aurantiaca	N	
O. dillenii	N	
O. ficus-indica	N	
O. leucantha	N	
O. rafinesquii	N?	
O. tortispina	N?	
Pereskia sp.	N	
Calycanthaceae — Laurales		
Calycanthus floridus	I	260, 292, 293, 191, 192
C. occidentalis	I	
Chimonanthus praecox	I	
Capparaceae — Capparales		
Capparis frondosa	I	294
Celastraceae — Celastrales		
Celastrus paniculata	I	295, 260, 29, 296, 197, 129
C. scandens	I	298, 258, 259, 88, 70-72, 143
Euonymus alata	I	255, 257, 145
E. alata f. *angustifolia*	I	
E. alata var. *apretus*	I	
E. americana	I	
E. dielsiana	I	
E. europaea	I	
E. fimbriata	I	
E. japonica	I	
E. latifolia	I	
E. macroptera	I	
E. maximowicziana	I	
E. miniata	I	
E. oxyphylla	I	
E. planipes	I	
E. sachalinensis	I	
E. vagans	I	
E. verrucosoides	I	
E. yedoensis	I	
E. koopmani	I	
Chenopodiaceae — Caryophyllales		
Beta lomatogona	N, Ap, Ag, P	119, 120, 299, 300
B. rubra	N, Ap, Ag, P	
B. trigyna	N, Ap, Ag, P	
B. vulgaris	N, Ap, Ag, P	

Taxa 1	Types and forms of apomixis 2	Ref. 3

Dicotyledones (continued)

Clusiaceae — Theales		
Clusia alba	N	301, 170, 17
Garcinia mangostana	N	
Combretaceae — Myrtales		
Combretum paniculatum	N	302, 303
C. pincianum	N	
Poivrea coccinea	N?	
Cucurbitaceae — Violales		
Momordica charantia	N	124, 380
Sicyos angulatus	N	
Dipterocarpaceae — Malvales		
Dipterocarpus castulatus	N	304
Shorea argentifolia	N	
S. agami	N	
S. macroptera	N	
S. ovalifolia spp. *sericera*	N	
S. resinosa	N	
S. usinosa	N	
Hopea odorata	N	
H. subalata	N	
Dipsacaceae — Dipsacales		
Scabiosa atropurpurea	N?	260
Euphorbiaceae — Euphorbiales		
Alchornea ilicifolia		
(*Coelebogune ilicifolia*)	N	28, 260, 29, 40, 107, 108
Euphorbia dulcis	N	305, 306, 102
E. platiphyllos	N	
E. rosea	N	
Mallotus japonica	N	
Fabaceae — Fabales		
Cassia artemisioides	N	307, 128, 249
C. australis var. *revoluta*	N	
C. circinata	N	
C. desolata var. *involucrata*	N	
C. eremophilla	I?, Ag	
C. nemophylla var. *coriacea*	N	
C. phyllodinea	N	
C. sturtii	N	
Milletia ovalifolia	N	
Gentianaceae — Gentianales		
Gentiana livonica	N?	283
G. lutea	N?	
G. punctata	N?	
Grossulariaceae — Saxifragales		
Ribes nigrum	N	308, 309
R. rubrum	N	
R. biebersteinii	N	

Taxa 1	Types and forms of apomixis 2	Ref. 3

Dicotyledones (continued)

Malpighiaceae — Polygalales		
Aspicarpa longipes	N	260, 189, 190, 310, 311, 312
Banisteria laurifolia	N	220, 67, 313
Heteropteris coerulea	N	
H. syringaefolia	N	
Hiptage madablota	N	
Malpighia sp.	N	
Stigmaphyllon emarginatum	N	
Thryallis glauca	N	
Meliaceae - Rutales		
Aphanamixis polystachya	N	105, 314
(*Amoora polystachya*)		
Lansium domesticum	N	
Moraceae — Urticales		
Ficus afganistanica	N	315, 317
F. carica × *Morus alba*	N	
F. roxburghii	N	
Streblus taxoides	N	
Myrsinaceae — Primulales		
Ardisia coriaria	N?	260, 201, 318
A. crispa	N?	
A. humilis	N?	
A. japonica	N?	
A. polytoca	N?	
A. serrulata	N?	
Myrtaceae - Myrtales		
Callistemon lanceolatus	N	
Eugenia cauliflora	N	42, 217, 48, 319, 320, 277, 10
(*Syzygium cauliflora*)		
E. bracteata	N	114, 321, 278, 277, 322, 323
E. cumingii	I	
E. formosa	I?	
E. fruticosa	I	
E. heyneana	I, N?	
E. hookeri	I	
E. jambos	I, N?	
E. jambolana	I	
E. malaccensis	I, Ap	
E. ugni	I	
Zyzygium (Eugenia) cumini	I, Ap	
Ochnaceae — Ochnales		
Ochna atropurpurea	I, Ap, Ag, P	325, 229, 230, 326, 327
O. multiflora	I, Ap, Ag, P	
O. serrulata	I, Ap, Ag, P	
Onagraceae — Myrtales		
Clarkia elegans	N	175, 135, 328, 329, 320, 330

Taxa 1	Types and forms of apomixis 2	Ref. 3

Dicotyledones (continued)

Epilobium hirsutum ×	N	
E. dodanaei		
E. angustifolium ×	N	
E. montanum		
E. dulcis	N	
Oenothera lamarckiana	N	
O. miricata × *O. biennis*		
Paeoniaceae — Paeoniales		
Paeonia caucasica	N, Ag	331
P. majko	N	
P. tenuifolia	N	
Polygonaceae — Polygonales		
Atraphaxis frutescens	N	133
Ranunculaceae — Ranunculales		
Caltha palustris	N?	332, 333
Ranunculus auricomus	N?	
Rosaceae — Rosales		
Alchemilla acutangula	N	260, 39, 166, 334, 118, 335
A. alpestris	N?, Ag	130, 336, 264
A. alpina	N?	
A. pastoralis	N?, Ag	
A. subcrenata	N, Ag	
Malus (cultivars)	N, P, AP	
Potentilla aurea	N, Ag	
P. geoides	N	
P. nepalensis ×	N, Ap	
P. argyrophylla		
P. reptans	N	
Pyrus (cultivars)	N, Ap, Ag	
Rutaceae — Rutales		
Aegle marmelos	N	260, 29, 43, 44, 223, 224, 100
Citrus aurantifolia	N	161, 222, 162, 207, 252, 337, 338
C. aurantium	N	163, 339–342, 83, 343, 277, 278
C. decuminata	N	344, 65, 98, 225, 345, 203, 346
C. hybridus	N	347, 164, 348, 349, 103, 104, 350
C. limon (cultivars)	N, Ap	220, 149, 351, 352, 219, 353, 354
C. macrocarpa	N	234, 355, 85, 86, 121, 185, 186
C. maxima	N	381, 122, 77, 78, 356, 211, 324
C. medica	N	148, 357, 358, 194
C. natsudaidai	N	
C. nobilis	N	
C. paradisi	N	
C. reticulata	N	
C. sinensis	N	
C. vulgaris	N	
C. juko	N	
C. unshiu	N	
C. medica × *C. limon*	N	

Taxa 1	Types and forms of apomixis 2	Ref. 3
Dicotyledones (continued)		
C. natsudaidai × *C. unshiu*	N	
Esenbeckia castanocarpa	N?	
E. jussueni	N?	
Feronia limonia	N	
Fortunella margarita	N	
Murraya exotica	N	
M. koenigii	N	
M. paniculata	N	
Poncirus trifoliata	N	
Ptelea trifoliata	N	
Ruta patavina	N	
Triphasia aurantiaca	N, Ap	
Zanthoxylum alatum	N, Ap	
Z. americanum	N	
Z. bungei	N	
Z. piperitum	N	
Z. planispinum	N	
Z. simulans	N	
Salicaceae — Salicales		
Salix multinervis	N, P	359
Solanaceae — Scrophulariales		
Lycopersicum sp.	N	175, 134, 135, 360, 201, 361
Petunia nyctaginiflora	N	
P. violacea	N	
Scopolia carniolica	I	
S. lurida	I	
Withania somnifera	N	
Simmondsiaceae — Hamamelidales		
Simmondsia chinensis	N?	362, 363
Symplocaceae — Ebenales		
Symplocos klotzschii	N	364
Thymelaeaceae — Thymelaeales		
Wikstroemia indica	N, Ag	30, 365
Tropaeolaceae — Geraniales		
Tropaeolum majus	N	260
Urticaceae — Urticales		
Elatostema acuminata	N, P, Ap	17
E. eurhynchum	N, P, Ap	
Violaceae — Violates		
Viola declinata	N?	223
Zygophyllaceae — Rutales		
Zygophyllum coccineum	N	367
Z. fabago	N	

Taxa 1	Types and forms of apomixis 2	Ref. 3

Monocotyledones

Agavaceae — Liliales		
Hosta (Funkia) coerulea	N	260, 19, 139
H. ovata	N	
Alliaceae — Liliales		
Allium nutans	I, Ag	261, 260, 29, 33, 34, 46, 47
A. odorum	I, Ag, Ap	188, 368, 106
A. roseum	I, Ag	
A. senescens	I, Ag	
Nothoscordum fragrans	I	
Amaryllidaceae — Liliales		
Calostemma purpureum	N?	228
Araceae — Arales		
Spathiphyllum patina	N	206, 48
Colchicaceae — Liliales		
Colchicum autumnale	N	260
Hemerocallidaceae — Liliales		
Hemerocallis mexicana	N	260
H. repanda	N?	
Liliaceae — Liliales		
Erythronium americanum	N	260, 29, 32, 135, 369
Tulipa gesneriana	N	
Ornithogalum umbellatum	N	
Orchidaceae — Orchidales		
Cynosorchis lilacina	N	29, 38, 370, 115, 371, 372, 116, 373
Cynorchis ampillaceae	N	263, 138, 17, 374, 202, 375
Epidendrum nocturnum	N, I?	
Gastrodoa elata	N, I?	
Gymnadenia conopsea	N?	
Maxillaria brasiliensis	N?	
M. cleistogama	N?	
Nigritella nigra	N	
Spiranthes cernua	I, Ap	
S. australis		
Zeuxina sulcata	N	
Z. strateumatica	N	
Zygopelatum mackayi	I, Ag	
Poaceae — Poales		
Agropyron sp.	N?	
Bothriochloa ischaemum	N, Ap	183, 176, 177, 140, 178-181,
Bromus inermis	N	269, 376
Cenchrus ciliaris	N	
C. glaucus	N	
Oryza sativa (cultivars)		
Pennisetum mezianum	N	

Taxa 1	Types and forms of apomixis 2	Ref. 3
Monocotyledones (continued)		
Poa pratensis	N, Ap, Ag, P	
Sorghum bicolor	N, Ap	
Smilacaceae — Smilacales		
Smilacina racemosa	N	45, 126
Tecophilaeaceae — Liliales		
Cyanella capensis	N	125
Trilliaceae — Smilacales		
Trillium erectum	N	377, 378, 93, 379
T. grandiflorum	N	
T. undulatum	N	

REFERENCES*

1. Schnarf, K., Embryologie der Angiospermen, Berlin, 1929, 417.
2. Maheshwari, P., *Embryology of Angiosperms*, New York, 1950, 467.
3. Nygren, A., Apomixis in the angiosperms, in *Handbuch der Pflanzenphysiologie*, Berlin, 1967, 551.
4. Modilewsky, Ya. S., Apomixis in angiosperms, *BOTZA9*, 5(2), 5, 1948.
5. Modilewsky, Ya. S., *Cytoembryology of Higher Plants*, Naukova Dumka, Kiev, 1963, 371.
6. Rutishauser, A., Fortpflanzungsmodus und Meiose apomiktischer Blütenpflanzen, *PRTOAF*, 6,3, 1967.
7. Poddubnaja-Arnoldi, V. A., *Cytoembryology of Angiosperms*, Nauka, Moscow, 1976, 507.
8. Lakchmanan, K. K. and Ambegaokar, K. B., Polyembryony, in *Embryology of Angiosperms*, Springer-Verlag, Berlin, 1984, 445.
9. Petrov, D. F., *Apomixis in Nature and in Experiment*, Nauka, Novosibirsk, 1988, 213.
10. Stebbins, G. L., Apomixis in angiosperms, *BOREA4*, 7(10), 507, 1941.
11. Battaglia, E., Apomixis, in *Recent Advances in the Embryology of Angiosperms*, Delhi, 1963, 221.
12. Grant, V., *Plant Speciation*, Mir, Moscow, 1984, 528.
13. Batygina, T. B., Some questions of embryology of plants, in *General Embryology*, Tokin, B. P., Ed., Vysshaja Shkola, Moscow, 1987, 470.
14. Winkler, H., Fortpflanzung der Gewächse, Apomixis, in *Handwörterbuch der Naturwissenschaften*, 2nd ed., Jena, 1934, 4.
15. Batygina, T. B., New approach to the system of reproduction in flowering plants, *ANL*, 1, 52, 1989.
16. Ernst, A., *Bastardierung als Ursache der Apogamie im Pflanzenreich. Eine Hypothese zur experimentellen Vererbungs und Abstammungslehre*, Jena, 1918, 666.
17. Poddubnaja-Arnoldi, V. A., *General Embryology of Angiosperms*, Nauka, Moscow, 1964, 482.
18. Fagerlind, P., Die Terminology der Apomixis Prozesse, *HEREAY*, 26(1), 1, 1940.
19. Fagerlind, F., Hormonale Substanzen als Ursache der Frucht und Embryobildung bei pseudogamen Hosta-biotypen, *SBOTAS*, 40, 230, 1946.
20. Gustafsson, A., Apomixis in higher plants. The mechanism of apomixis, *LUARAS*, 2, 42, 3, 1, 1946.
21. Gustafsson, A., Apomixis in higher plants. The casual aspects of apomixis, *LUARAS*, 2, 43, 2, 71, 1947.
22. Gustafsson, A., Apomixis in higher plants. Biotype and species formation, *LUARAS*, 2, 43, 12, 181, 1947.
23. Khokhlov, S. S., Classification of apomixis of angiosperms, *DASBAQ*, 119, 4, 812, 1958.
24. Petrov, D. F., *Genetically Controlled Apomixis*, Nauka, Novosibirsk, 1964, 187.
25. Petrov, D. F., *Genetic Fundamentals of Apomixis*, Nauka, Novosibirsk, 1979, 276.
26. Solntzeva, M. P., Embryological classification of apomixis in angiosperms, *GNTCAM*, V, 8, 20, 1969.
27. Solntzeva, M. P., Classification of embryo sacs of apomictic plants, *ANL*, 1, 4, 1990.
28. Smith, J., Notice of a plant which produces seeds without any apparent action of pollen, *Trans. Linn. Soc. (London)*, 18, 509, 1841.
29. Strasburger, E., Uber Polyembryonie, *ZNFCAP, N. F.*, 4, 1, 1878.
30. Winkler, H., Über Parthenogenesis bei *Wikstroemia indica* (L.) C.A.Mey., *Ann. Jard. Bot. Buitenz.*, 2, 5(20), 208, 1906.
31. Winkler, H., Parthenogenesis und Apogamie im Pflanzenreiche, *Progr. Bot. (Jena)*, 2, 3, 1, 1908.
32. Jeffrey, E. C., Polyembryony in *Erythronium americanum*, *ANBOA4*, 9, 537, 1895.

* Many of the references in this listing are identified using the CODEN system, a unique serial identifier consisting of six characters. It is assigned and administered by International CODEN Service, Chemical Abstracts Service, Columbus, Ohio.

Apomixis in Angiosperms

33. Tretjakow, S., Die Betheiligung der Antipoden in Fällen der Polyembryony bei *Allium odorum, BEDBAP,* 13, 13, 1895.
34. Hegelmaier, F., Zur Kenntnis der Polyembryonie von *Allium odorum* L., *Bot. Ztg.,* 1, 133, 1897.
35. Ganong, W. F., Upon polyembryony and its morphology in *Opuntia vulgaris, BOGAA5,* 25, 4, 221, 1898.
36. Montemartini, L., Contributo allo studio dell'anatomia del frutto e del seme delle *Opunzie, AIBBAW,* 5, 59, 1899.
37. Hull, E. D., Polyembryony in *Opuntia rafinesquii, ANBOA4,* 21, 2, 56, 1915.
38. Leavitt, R. G., Notes on the embryology of some New England Orchids, *RHODAB,* 3, 7, 61, 1901.
39. Murbeck, S., Über normalien im Baue des Nucellus und des Embryo-sackes bei parthenogenetischen Arten der Gattung *Alchemilla, Acta Reg. Soc. Physiogr.,* 13, 2, 10, 1902.
40. Hegelmaier, F., Zur Kenntnis der Polyembryonie von *Euphorbia dulcis* Jacg., *BEDBAP,* 21, 6, 1903.
41. Furlani, J., Zur Embryologie von *Colchicum autumnale* L., *OBOZA2,* 54, 373, 1904.
42. Cook, M. T., Notes on polyembryony, *TORRA3,* 7, 113, 1907.
43. Longo, B., La poliembrionia nello *Xanthoxylum* bungei senza fecondation, *Bull. Soc. Bot. Ital.,* 11, 113, 1908.
44. Osawa, J., Cytological and experimental studies in *Citrus, J. Imper. Univ. Tokyo, Coll. Agr.,* 4, ,1, 83, 1912.
45. McAllister, F., On the cytology and embryology of *Smilacina racemosa, TWASAB,* 17, 599, 1913.
46. Modilewsky, Ya. S., Investigation of polyembryony in *Allium odorum, Vestnik Kiev. Bot. Sada,* 2, 9, 1925.
47. Modilewsky, Ya. S., Embryo formation in *Allium odorum, Vestnik Kiev. Bot. Sada,* 12, 45, 1925.
48. van der Pijl, L., Über die Polyembryonie bei *Eugenia, RTBNAQ,* 31, 113, 1934.
49. Webber, J. M., Polyembryony, *BOREA4,* VI, 11, 575, 1940.
50. Schnarf, K., *Vergleichende Embryologie der Angiospermen.* Berlin, 1931, 354.
51. Stebbins, G. L., *Variation and Evolution in Plants,* Columbia University Press, New York, 1950, 643.
52. Asker, S., Gametophytic apomixis: elements and genetic regulation, *HEREAY,* 93, 277, 1980.
53. Nogler, G. A., Gametophytic apomixis, in *Embryology of Angiosperms,* Springer-Verlag, Berlin, 1984, 475.
54. Levina, R. E., On evolutionary basis of apomixis in angiosperms, *TMPIAK,* 27, 2, 70, 1972.
55. Khokhlov, S. S., Apomixis: classification and distribution in angiosperms, *USSGAE,* 1, 43, 1967.
56. Khokhlov, S. S., Evolutionary and genetic problems of apomixis in angiosperms, in *Apomixis and Selection,* Nauka, Moscow, 1970, 7.
57. Khokhlov, S. S., Zaitzeva, M. I., and Kuprijanov, P. G., *The Identification of Apomictic Forms in the Flora of Flowering Plants of the USSR,* Izd. Sarat. University, Saratov, 1978, 224.
58. Archibald, E. A., The development of the ovule and seed of jointed cactus (*Opuntia aurantiaca* Lindley), *SAJSAR,* 36, 195, 1939.
59. Nygren, A., Apomixis in angiosperms, *BOREA4,* 20, 10, 577, 1954.
60. Maheshwari, P., Ed., *Recent Advances in the Embryology of Angiosperms,* Delhi, 1963, 467.
61. Maheshwari, P. and Sachar, R. C., Polyembryony, in *Recent Advances in the Embryology of Angiosperms,* Delhi, 1963, 265.
62. Naumova, T. N., Ultrastructural aspects of apomixis, in *Proc. XI Int. Symp. Embryology and Seed Reproduction,* Nauka Leningrad, 1992, 393.

63. Naumova, T. N., Nucellar and Integumentary Embryony in the System of Seed Reproduction of Flowering Plants, doctoral dissertation, Leningrad, 1991, 626.
64. Lebegue, A., La polyembryonie chez les Angiosperms, *BSBFAN*, 99, 7, 329, 1952.
65. Yakovlev, M. S., Polyembryony in higher plants, in *Problems of Current Embryology*, Izd. Leningrad University, Leningrad, 1956, 35.
66. Yakovlev, M. S., Polyembryony in higher plants and principles of its classification, *PHYMAW*, 17, 1–4, 278, 1967–1968.
67. Poddubnaja-Arnoldi, V. A., *The Characteristic of the Families of Angiosperms by Cytoembryological Features*, Nauka, Moscow, 1982, 352.
68. Naumova, T. N. and Yakovlev, M. S., Adventive embryony in angiosperms, *BOTZA9*, 57, 8, 1006, 1972.
69. Rieger, R., Michaelis, A., and Green, M., *Glossary of Genetics and Cytogenetics*, Gustav Fisher, Iena, 1976, 647.
70. Naumova, T. N., The development of sexual and apomictic embryos in *Euonymus europaea* L., *BOTZA9*, 55, 5, 638, 1970.
71. Naumova, T. N., Polyembryony in *Euonymus macroptera* Rupr. and *E. planipes* (Koehne) Koehne, *BOTZA9*, 55, 9, 1270, 1970.
72. Naumova, T. N., Adventive Embryony in Genus *Euonymus* (Celastraceae), Ph.D. dissertation, Leningrad, 1971, 117.
73. Naumova, T. N., Nucellar polyembryony in the genus *Sarcococca* (Buxaceae), *BOTZA9*, 65, 2, 230, 1980.
74. Naumova, T. N., On the embryology of representatives of the family Buxaceae, *BOTZA9*, 66, 8, 1135, 1981.
75. Naumova, T. N., Family Cactaceae, in *Comparative Embryology of Flowering Plants (Phytolaccaceae-Thymelaeaceae)*, Nauka, Leningrad, 1983, 19.
76. Naumova, T. N., Adventive embryony in angiosperms, in *Thesis of XII International Botanical Congress*, Nauka, Leningrad, 1975, 227.
77. Wakana, A., and Uemoto, S., Adventive embryogenesis in *Citrus*. I. The occurrence of adventive embryos without pollination and fertilization, *AJBOAA*, 74, 4, 517, 1987.
78. Wakana, A. and Uemoto, S., Adventive embryogenesis in *Citrus* (Rutaceae). II. Postfertilization development, *AJBOAA*, 75, 7, 1033, 1988.
79. Johri, B. M., Ed., *Embryology of Angiosperms*, Springer-Verlag, Berlin, 1984, 830.
80. Batygina, T. B. and Butenko, R. G., Morphological potentiality of the embryos of angiosperms (on the example of representatives of genus *Paeonia* family Paeoniaceae), *BOTZA9*, 66, 11, 1531, 1981.
81. Naumova, T. N. and Yakovlev, M. S., Sexual reproduction of *Opuntia ficus-indica* (Cactaceae), *BOTZA9*, 63, 4, 548, 1978.
82. Reynolds, B. S., The use of lead citrate at high PH as an electron-opaque stain in electron microscopy, *JCLBA3*, 17, 208, 1963.
83. Webber, H. J. and Batchelor, L. D., Eds., *The Citrus Industry: History, Botany and Breeding*, University of California Press, Berkeley, 1948, 933.
84. Milyaeva, E. L., Electron microscope investigation of exine formation in microspores of *Citrus sinensis*, *DASBAQ*, 168, 2, 467, 1966.
85. Esen, A. and Soost, R., Precocious development and determination of spontaneous triploid seeds in *Citrus*, *HDTYAT*, 64, 3, 147, 1973.
86. Esen, A. and Soost, R. K., Adventive embryogenesis in *Citrus* and its relation to pollination and fertilization, *AJBOAA*, 64, 6, 607, 1977.
87. Merzlikina, T. I., On the biology of *Euonymus japonica*, *BYGBAA*, 19, 135, 1954.
88. Copeland, H. F., Morphology and embryology of *Euonymus japonica*, *PHYMAW*, 16, 3, 326, 1966.
89. Wiger, J., Ein neuer Fall von autonomer Nuzellarpolyembryonie, *BONOAE*, 368, 1930.
90. Wiger, J., Embryological Studies on the Families Buxaceae, Meliaceae, Simarubaceae and Burseraceae, Thesis, Lund, 1935, 129.

91. Wiger, J., Reply to remarks on my paper on Buxaceae. Meliaceae etc., *BONOAE*, 6, 585, 1936.
92. Kordyum, E. L., Microsporogenesis and the peculiarity of tapetum formation in some species of genus *Vincetoxicum* Moench., *UKBZAW*, 18, 5, 3, 1961.
93. Naumova, T. N., Peculiarity of macrogametogenesis and post-fertilization development of *Trillium camschatcense* Ker-Gawl., *Soc. Bot. France, Actual. Bot.*, 1–2, 183, 1978.
94. Tiagi, V. D., Studies in the floral morphology of *Opuntia dillenii* Haworth., *BONOAE*, 4, 343, 1954.
95. Chopra, R. N., The mode of embryo sac development in *Opuntia aurantiaca* Lindl. A reinvestigation, *PHYMAW7*, 3–4, 403, 1957.
96. Maheshwari, P. and Chopra, R. N., Polyembryony in *Opuntia dillenii*, *CUSCAM*, 32, 4, 130, 1954.
97. Maheshwari, P., and Chopra, R. N., The structure and development of the ovule and seed of *Opuntia dillenii* Haw., *PHYMAW*, 5, 1, 112, 1955.
98. Johri, B. M. and Ahuja, M. R., Development of endosperm and nucellar polyembryony in *Aegle marmelos* Correa., *CUSCAM*, 25, 5, 162, 1956.
99. Johri, B. M., and Ahuja, M. R., A contribution to the floral morphology and embryology of *Aegle marmelos*, *PHYMAW*, 7, 10, 1957.
100. Frost, H. B., Seed reproduction: development of gametes and embryos, in *Citrus Industry*, Vol. 1, University of California Press, Berkeley, 1948, 767.
101. Roy, S. K., Embryology of *Eugenia jambos* L., *CUSCAM*, 22, 8, 249, 1953.
102. Kapil, R. N., Some embryological aspects of *Euphorbia dulcis* L., *PHYMAW*, 11, 1, 24, 1961.
103. Desai, S., Cytology and embryology of Rutaceae, *PHYMAW*, 12, 2, 178, 1962.
104. Desai, S., Polyembryony in *Xanthoxylum* Mill., *PHYMAW*, 12, 2, 184, 1962.
105. Ghosh, R. B., A note on the nucellar polyembryony in *Aphanamixis polystachya* (Wall.)Parker., *CUSCAM*, 1, 4, 165, 1962.
106. Hakansson, A., Die Samenbildung bei *Nothoscordum fragrans*, *BONOAE*, 1, 129, 1953.
107. Carano, E., Sul particolare sviluppo del gametofito femminile di *Euphorbia dulcis* L., *Rend. R. Accad. Lincei*, 6a, 1, 633, 1925.
108. Carano, E., Ulteriori osservazoni su *Euphorbia dulcis* L. in rapporto col suo comportamento apomitto, *ANBOA4*, 17, 50, 1926.
109. Rodkiewicz, B., Megasporogenesis in angiosperms (a retrospect). I. Callose distribution, *BLTBAI*, 23, 2, 109, 1981.
110. Kapil, R. N., and Bhatnagar, A. K., Ultrastructure and biology of female gametophyte in flowering plants, in *International Review of Cytology*, New York, 1981, 70, 291.
111. Willemse, M. T. M. and Franssen-Verheijen, M. A. W., Cell organelle changes during megasporogenesis and megagametogenesis in *Gasteria verrucosa* (Mill.) Haw., *Soc. Bot. France, Actual Bot.*, 125, 1–2, 187, 1978.
112. Willemse, M. T. M., and Bednara, J., Polarity during megasporogenesis in *Gasteria verrucosa*, *PHYMAW*, 29, 2, 156, 1079.
113. Rodkiewicz, B., Megasporogenesis in angiosperms (a retrospect). II. Polarization phenomena, *BLTBAI*, 23, 2, 115, 1981.
114. Roy, S. K., Embryology of *Eugenia malaccensis* Lam., *CUSCAM*, 29, 5, 189, 1960.
115. Afzelius, K., Die Embryobildung bei *Nigrittella nigra*, *SBOTAS*, 22, 82, 1928.
116. Swamy, B. G. L., The embryology of *Zeuxine sulcata* Lindley., *NEPHV*, 45, 132, 1946.
117. Vosito Sinoto, On the nuclear division and partial sterility in *Oenothera lamarckiana* Ser., *Bot. Mag. Tokyo Bot. Soc.*, 36, 92, 1932.
118. Gentscheff, G., Über die pseudogame Fortpflanzung bei *Potentilla*, *GENEA3*, 20, 388, 1938.
119. Jassem, B., Embryology and genetics of apomixis in the section Corollinae of the genus *Beta*, *ABCBAM*, 19, 149, 1976.
120. Jassem, B., Apomixis in the genus *Beta*, *ANL*, 2, 7, 1990.
121. Esen, A., Soost, R. K., and Geraci, G., Genetic evidence for the origin of diploid megagametophytes in *Citrus*, *HDTYAT*, 70, 1, 5, 1979.

122. Wakana, A., Iwamasa, M., and Uemoto, S., Seed development in relation to ploidy of zigotic embryo and endosperm in polyembryonic *Citrus*, *Proc. Int. Soc. Citric.*, 2, 35, 1981.

123. Baker, H. G., Apomixis and polyembryony in *Pachira oleaginea* (Bombacaceae), *AJBOAA*, 47, 4, 296, 1960.

124. Agrawal, J. S. and Singh, S. P., Nucellar polyembryony in *Momordica charantia* Lirin., *SCINAL.* 22, 11, 630, 1957.

125. Vos, M. P., Seed development in *Cyanella capensis* L.: a case of polyembryony, *SAJSAR*, 46, 220, 1950.

126. Gorham, A. L., The question of fertilization in *Smilacina racemosa* (L.) Desf., *PHYMAW*, 3, 12, 44, 1953.

127. Sachar, R. C., and Chopra, R. N., A study of the endosperm and embryo in *Mangifera* L., *IJASA3*, 27, 2, 219, 1957.

128. Niranjan, P., Development of the seed of *Millettia ovalifolia*, *BOGAA5*, 122, 2, 130, 1960.

129. Andersson, A., Studien über die Embryologie der Familien Celastraceae, Oleaceae und Apocynaceae, *LUARAS*, 27, 7, 1, 1931.

130. Grevtzova, N. A., Apomixis in genus *Malus* Mill. and *Pyrus* L., in *Apomixis and Cytoembryology*, Isd. Sarat. University, Saratov, 1978, 4, 22.

131. Naumova, T. N., den Nijs, A. P. M., and Willemse, M. T. M., Cytological approach to characterize apomixis in *Poa pratensis* genotypes, *ANL*, 4, 31, 1992,

132. Zueva, G. V., and Egorova, N. N., Fertilization in *Phleum stepposum*, in *Ontogenesis of Herbal and Polycarpic Plants*, Sverdlovsk, 1980, 53.

133. Edman G., Apomeiosis und Apomixis bei *Atraphaxis frutescens* C. Koch., *Acta Horti. Berg.*, 11, 13, 1931.

134. Haberlandt, G., *Die Vorstufen und Ursachen der Adventivembryonie*, *Sitz. Preuss. Acad. Wiss.*, 386, 1922.

135. Haberlandt, G., Über Zellteilungshormone und ihre Beziehungen zur Wundheilung, Befruchtung, Parthenogenesis und Adventivembryonie, *BIZNAT*, V, 42, 145, 1922.

136. Kandelaki, G. V., Pathways of apomictic endosperm development, *SAKNAH*, 36, 2, 423, 1964.

137. Erdelska, O., Dynamics of the development of embryo and endosperm *(Papaver somniferum, Nicotiana tabacum, Jasione montana)*, *BLOAAO*, 40, 1, 17, 1985.

138. Heslop-Harrison, J., The physiology of reproduction in *Dactylorchis*, *BONOAE*, 110, 1, 28, 1957.

139. Hu, S. V., Studies in the polyembryony of *Hosta coerulea* Tratt., *Acta Bot. Sinica*, 11, 1, 16, 1963.

140. Shanthamma, C. and Narayan, K. N., Formation of nucellar embryos with total absence of embryo sacs in two species of Gramineae, *ANBOA4*, 41, 172, 469, 1977.

141. Shanthamma, C., Apomixis in *Cenchrus glaucus* Mudaliar et Sundaraj, *PIACAP*, 91, 1, 25, 1982.

142. Maheshwari, P. and Ranga Swamy, N. S., Polyembryony and *in vitro* culture of embryos of *Citrus* and *Mangifera*, *IJHOAQ*, 15, 275, 1958.

143. Naumova, T. N., The fine structure of inner integument of *Euonymus macroptera* in respect of integumentary embryos formation, in *Abstracts IV Vsesouz. Symp. Ultrastructure of plant cells*, Zinate, Riga, 1978, 180.

144. Naumova, T. N. and Willemse, M. T. M., Nucellar polyembryony in *Sarcococca humilis*. Ultrastructural aspects, *PHYMAW*, 32, 1, 94, 1982.

145. Naumova, T. N. and Willemse, M. T. M., Nucellar and integumentary polyembryony in some angiosperms: ultrastructural aspects, in *Proc. VII Int. Symp.on Plant Embryology*, High Tatra, Bratislava, 1982, 357.

146. Naumova, T. N. and Willemse, M. T. M., Ultrastructural aspects of nucellar polyembryony of *Sarcococcu humilis* (Buxaceae). Nucellus and initial cells of nucellar embryos, *BOTZA9*, 68, 8, 1044, 1983.

147. Naumova, T. N. and Willemse, M. T. M., Ultrastructural aspects of nucellar polyembryony of *Sarcococca humilis* (Buxaceae). Initial cells division and nucellar embryos development, *BOTZA9*, 68, 9, 1184, 1983.

148. Wilms, H. J., van Went, J. L., Cresti, M., and Ciampolini, F., Adventive embryogenesis in *Citrus, CARYAB*, 36, 1, 65, 1983.

149. Polunina, N. N., The localization of ascorbic acid at nucellar embryony of *Citrus, BYGBAAR*, 58, 65, 1965.

150. Robards, A. W., Plasmodesmata, *ARPHAD*, 26, 13, 1975.

151. Gamalei, Yu. V., Plasmodesmata: intercellular communication in plants, *FZRSAV*, 32, 1, 176, 1985.

152. Dickinson, H. G. and Heslop-Harrison, J., Ribosomes, membranes and organelles during meiosis in angiosperms, *Philipp. Trans. R. Soc. (London)*, 277, 955, 327, 1977.

153. Dickinson, H. G., Cytoplasmic differentiation during microsporogenesis in higher plants, *ASBNA2*, 50, 1–2, 3, 1981.

154. Medina, F. J., Risueno, M. C., and Rodrigues-Garcia, M. J., Evolution of the cytoplasmic organelles during meiosis in *Pisum sativum* L., *PLANAB*, 151, 215, 1981.

155. Gabaraeva, N. I., The development of spores in *Psilotum nudum* (Psilotaceae). The changes of cytoplasm and organelles of spore mother cells in the premeiotic interphase — to leptotene prophase of the meiosis, *BOTZA9*, 69, 11, 1441, 1984.

156. Gabaraeva, N. I., The development of spores in *Psilotum nudum* (Psilotaceae): the changes in cytoplasm and organelles of spore mother cells from zygotene to pachytene, *BOTZA9*, 69, 12, 1612, 1984.

157. Gabaraeva, N. I., The development of spores in *Psilotum nudum* (Psilotaceae): changes in cytoplasm and organelles of spore mother cells in metaphase and telophase I of miosis, *BOTZA9*, 70, 4, 441, 1985.

158. De Boer-de-Jeu, M. J., Ultrastructural aspects of megasporogenesis and initiation of megagametogenesis in *Lilium, BSBFAN*, 125, 1–2, 125, 1978.

159. De Boer-de-Jeu, M. J., *Megasporogenesis. A Comparative Study of the Ultrastructural Aspects of Megasporogenesis in Lilium, Allium and Impatiens*, Meded. Landbouwhogesch, Wageningen, 1978, 16, 128.

160. Willemse, M. T. M., Polarity during megasporogenesis and megagametogenesis, *PHYMAW*, 31, 2, 124, 1981.

161. Nagai, K. and Tanikawa, T., On *Citrus* pollination, in *Proc. Third Pan-Pacific Science Congress*, Tokyo, 11, 2023, 1926.

162. Toxopeus, H. J., De polyembryonie van *Citrus* en haar betekenis voor de cultur, *Vereen Landbouw Nederl. Indië Landbouw Tidschr.*, 6, 391, 1930.

163. Wong, C. J., The influence of pollination on seed development in certain varieties of *Citrus, PASHA6*, 37, 161, 1940.

164. Kapanadze, I. S., Fertilization and additional embryos formation in *Citrus, Tr. Sukhum. Opytn. Stan. Subtrop. Kult.*, 1, 189, 1967.

165. Juliano, J. B. and Cuevas, N. L., Floral morphology of the Mango (*Magnifera indica* Linn.) with special reference to the Pico variety from the Philippines, *PHAGAU*, 21, 7, 449, 1932.

166. Popoff, A., Über die Fortpflanzungsverhältnisse der Gattung *Potentilla, PLANAB*, 24, 510, 1935.

167. Kordyum, E. L., Investigation of the polyembryony in *Vincetoxicum officinale* Moench., *UKBZAW*, 18, 3, 48, 1961.

168. Wright, T., Pollination and the seediness of Marsh grapefruit, *Trinidad Tobago Agric. Soc. Proc.*, 37, 51, 1937.

169. Woodworth, R. H., Parthenogenesis and polyembryony in *Alnus rugosa* (Duroi), *SCIEAS*, 70, 192, 1929.

170. Woodworth, R. H., Cytological studies in the Betulaceae. III. Parthenogenesis and polyembryony in *Alnus rugosa, BOGAA5*, 39, 108, 1930.

171. Horn, C. L., Existence of only one variety of cultivated mangosteen explained by asexually formed seeds, *SCIEAS*, 92, 237, 1940.

172. Horn, C. L., The frequency of polyembryony in twenty varieties of Mango, *PASHA6*, 42, 318, 1943.

173. Yang-Hsu-Yen, Fertilization and development of embryo on Satsuma orange (*Citrus unshiu* Marc.) and Natsudaidai (*C. natsudaidai* Hayata), *J. Jpn. Soc. Hort. Sci.*, 37, 2, 102, 1968.

174. Duncan, E. J., Ovule and embryo ontogenesis in *Bambacopsis glabra* (Pasq.) A. Robins., *ANBOA4*, 37, 136, 677, 1970.

175. Haberlandt, G., *Über experimentelle Erzeugung von Adventivembryonen bei Oenothera lamarckiana*, Sitzungsber. Preuss, Acad. Wiss., 1921, 695.

176. Dewey, D. R., Polyembryony in Agropyron, *CRPSAY*, 4, 3, 313, 1964.

177. Schertz, K. F. and Bashow, E. C., Apospory in *Sorghum bicolor* (L.) Moench., *SCIEAS*, 170, 3955, 338, 1970.

178. Moskova, R. D., An electron microscopic study of the nucellus cells of *Botriochloa ischaemum* L., *CARYAB*, 28, 295, 1975.

179. Moskova, R. D., Apomixis in *Botriochloa ischaemum* L., in *Apomixis and its Utilization in Plant Breeding*, Kolos, Moscow, 1976, 101.

180. Moskova, R. D., Apomixis in *Botriochloa ischaemum* L., in *Apomixis and Cytoembryology of Plants*, Vol. 4, Izd. Sarat. University, Saratov, 1978, 81.

181. Batygina, T. B. and Freiberg, T. E., Polyembryony in *Poa pratensis*, *BOTAZA9*, 64, 1, 793, 1979.

182. Batygina, T. B., Nucellar embryoidogeny in *Poa pratensis*, *ANL*, 3, 18, 1991.

183. Nishimura, M., On the germination and the polyembryony of *Poa pratensis* L., *BOMZA8*, 36, 47, 1922.

184. Nishimura, M., Comparative morphology and development of *Poa pratensis*, *Phleum pratense* and *Setaria italica*, *JJBOA7*, 1, 55, 1922.

185. Kobayashi, S., Ikeda, I., and Nakatani, M., Studies on the nucellar embryogenesis in *Citrus*. I. Formation of nucellar embryo and development of ovule, *Bull. Fruit Tree Res. St. Jpn., Ser. E*, 2, 11, 1978.

186. Kobayashi, S., Ikeda, I., and Nakatani, M., Studies on nucellar embryogenesis in *Citrus*, *J. Jpn. Soc. Hort. Sci.*, 48, 2, 179, 1979.

187. Afzelius, K., Apomixis and polyembryony in *Zygopetalum mackayi* Hook., *Act. Not.*, 19, 2, 7, 1959.

188. Stenar, H., Studien über die Entwicklungsgeschichte von *Nothoscordum fragrans* Kunth. und *Nistriatum* Kunth., *SBOTAS*, 26, 25, 1932.

189. Subha Rao, A. M., A note on the development of the female gametophytes of some Malpighiaceae and polyembryony in *Hiptage madablota*, *CUSCAM*, 6, 6, 280, 1937.

190. Subba Rao, A. M., The ovule and embryo sac development of some Malpighiaceae, *CUSCAM*, 8, 79, 1939.

191. Brofferio, J., Osservazioni sullo sviluppo delle Calycanthaceae, *ANBOA4*, 18, 387, 1930.

192. Kamelina, O. P., Family Calycanthaceae, in *Comparative Embryology of Flowering Plants (Winteraceae-Juglandaceae)*, Nauka, Leningrad, 1981, 69.

193. Naumova, T. N., The peculiarity of nucellar tissue development and nucellar polyembryony in *Opuntia elata (Cactaceae)*, *BOTZA9*, 63, 3, 344, 1978.

194. Naumova, T. N., Family Rutaceae, in *Comparative Embryology of Flowering Plants (Brunelliaceae-Tremandraceae)*, Nauka, Leningrad, 1985, 131.

195. Jensen, W. A., Reproduction of flowering plant, in *Dynamic Aspects of Plant Ultrastructure*, Robards, A. W., Ed., McGraw-Hill, England, 1974, 481.

196. Jensen, W. A., The role of cell division in angiosperm embryology, in *Cell Division of Higher Plants*, London, 1976, 391.

197. Jensen, W. A., Comienzos de la embriogenesis en angiospermas, *KURTAK*, 14, 63, 1981.

198. Zhukova, G. Ya. and Savina, G. I., Electron-microscopic investigation of the embryo of *Epipactis atrorubens* (Hoffm.) Schulk., *BOTZA9*, 63, 9, 1241, 1978.

199. Ashley, T., Zygote shrinkage and subsequent development in some Hibiscus hybrids, *PLANAB*, 108, 4, 303, 1972.

200. Souèges, R., *Embryogènie et Classification. Essai d'un Système Embryogènique: Partie Gènèrale*, Paris, 1939, 95.
201. Johansen, D. A., *Plant Embryology*, Waltham, MA, 1950, 305.
202. Karanth, K. A., Swamy, B. G. L., and Arekal, G. D., Embryogenesis in sexual and asexual species of *Zeuxine* (Orchidaceae), *PIACAP*, 90, 1, 1, 1981.
203. Banerji, J. and Pal, S., Studies in the embryology of *Feronia limonia* Swingle, *BBSGAQ*, 12, 1–2, 18, 1958.
204. Lakshmanan, K. K., Irregular embryogeny in angiosperms, in *Seminar on Morphology, Anatomy and Embryology of Land Plants*, Delhi, 1969, 69.
205. Hausner, G., Embryogenesis und Nucellar-Polyembryonie bei *Cynanchum/Vincetoxicum* — Arten, *BEPFAT*, 52, 101, 1976.
206. Schurhoff, P. N. and Jussen, J., Nuzellarpolyembryonie bei *Spathiphyllum patinii* (Hoog.) N.E.Br., *BEDBAP*, 43, 454, 1925.
207. Chakravarty, R. S., Polyembryony in *Murrays koenigii*, *CUSCAM*, 3, 361, 1935.
208. Haccius, B. and Lakshmanan, K. K., Adventive Embryonen — embryoide-adventive Knospen. Ein Beitrag zur Klärung der Begriffe, *OBOZAZ*, 11, 1–5, 145, 1969.
209. Tiagi, V. D., Polyembryony in *Mammillaria tenuis* DC., *BBUSA4*, 6, 25, 1956.
210. Tiagi, V. D., Studies in floral morphology. III. A contribution to the floral morphology of *Mammillaria tenuis* D.C., *J. Univ. Saugar*, 6, 11, Sec. B, 7, 1957.
211. Bai, V. Narmatha, and Lakshmanan, K. K., Occurrence of nucellar polyembryony in *Toddalia asiatica* Lamk., *CUSCAM*, 51, 374, 1982.
212. Naumova, T. N., Embryos of angiosperms: origin and development, *ANNSA8*, 23, 161, 1988.
213. Bhatnagar, S. P., and Sawhney, V., Endosperm — its morphology, ultrastructure and histochemistry, in *International Review of Cytology*, New York, 1981, 73, 55.
214. Varsha Parikh, Suspensor in angiosperms, *Botanica (India)*, Silver Jubillie Volume, 1975, 84.
215. Nagl, W., Ultrastructural and developmental aspects of autolysis in embryo-suspensors, *BEDBAP*, 89, 2–3, 301, 1976.
216. Nagl, W., Early embryogenesis in *Tropaeolum majus* L.: ultrastructure of the embryo-suspensor, *BPPFA4*, 170, 3, 253, 1976.
217. Tiwary, N. K., On the occurrence of polyembryony in the genus *Eugenia*, *JIBSAC*, 5, 124, 1926.
218. Sahai, R. and Roy, S. K., Polyembryony in *Eugenia heyneana* Duthie., *SCINAL*, 28, 1, 37, 1962.
219. Chandra, Dh. and Shanker, G., Polyembryonic studies in certain indigenous species of *Citrus*, *SCINAL*, 31, 7, 373, 1965.
220. Singh, B. P. and Soule, M. J., Studies in polyembryony and seed germination of trifoliata orange *(Poncirus trifoliata)*, *IJHOAQ*, 20, 1, 21, 1963.
221. Selivanov, A. C., *Multiembryony of Seeds and Plant Breeding. 1. Prospects of Utilization and Creation of Multiembryonic Forms of Cultivated Plants*, Izd. Sarat. University, Saratov, 1983, 83.
222. Cappalletti, C., Sterilita di origine micotica nella *Ruta patavina* L., *ANBOA4*, 18, 145, 1929.
223. Frost, H. B., Polyembryony, heterozygosis and chimeras in *Citrus*, *HILGA4*, 1, 16, 365, 1926.
224. Frost, H. B., Nucellar embryony and juvenile character in clonal variaties of *Citrus*, *HDTYAT*, 29, 423, 1926.
225. Furusato, K., Ohta, V. and Ishibashi, K., Studies on polyembryony in *Citrus*, *SEZIA3*, 8, 40, 1957.
226. Juliano, J. B., Origin of embryos in the Strawberry Mango, *PYBTAK*, 54, 553, 1934.
227. Juliano, J. B., Embryos of Carabao Mango *(Magnifera indica* L.), *PHAGAU*, 25, 749, 1937.
228. Schlimbach, H., Beitrag zur Kenntnis der Samenanlagen und Samen der Amaryllidaceae mit Berücksichtigung des Wassergehalts der Samen, *FABZAZ*, 117, 41, 1924.
229. Chiarugi, A. and Francini, E., Apomissia in *Ochna serrulata*, *AISCAI*, 18, 1929.

230. Chiarugi, A. and Francini, E., Apomissia in *Ochna serrulata* Walp., *NUGBAC*, 37, 1, 1930.
231. Kordyum, E. L., *Evolutionary Cytoembryology in Angiosperms*, Naukova Dumka, Kiev, 1978, 219.
232. Halperin, W., Alternative morphogenetic events in cell suspensions, *AJBOAA*, 53, 443, 1966.
233. Halperin, W., Embryos from somatic plant cells, *Proc. Int. Symp. Soc. Cell Biol.*, 9, 169, 1970.
234. Button, J. and Bornman, C. H., Development of nucellar plants from unpollinated and unfertilized ovules of the Washington navel orange *in vitro*, *SAJSAR*, 37, 2, 127, 1971.
235. Button, J., Kochba, J., and Bornman, C. H., Fine structure and embryoid development from embryogenic ovular callus of Shamonti Orange (*Citrus sinensis* Osb.), *JEBOA6*, 25, 85, 446, 1974.
236. Raghavan, V., *Experimental Embryogenesis in Vascular Plants*, Academic Press, London, 1976, 603.
237. Batygina, T. B., Vasilyeva, V. E. and Mametjeva, T. B., The problems of morphogenesis *in vivo* and *in vitro*. Embryogenesis of angiosperms, *BOTZA9*, 63, 1, 87, 1978.
238. Johri, B. M. and Rao, P. S., Experimental embryology in *Embryology of Angiosperms*, Springer-Verlag, Berlin, 1984, 735.
239. Williams, E. G. and Maheshwaran, G., Somatic embryogenesis, *ANBOA4*, 57, 4, 443, 1986.
240. Thomas, E., The fine structure of the embryogenic callus of *Ranunculus sceleratus*, *JNCSAI*, 11, 95, 1972.
241. Tisserat, B. and De Mason, D. A., A histological study of development of adventive embryos in organ cultures of *Phoenix dactylifera* L., *ANBOA4*, 46, 4, 465, 1980.
242. Haccius, B., Haben "Gewebekulturen-embryonen" einen Suspensor?, *BEDBAP*, 78, 11, 10, 1965.
243. Swamy, B. G. L. and and Krishnamurthy, K. V., On embryos and embryoides, *PIACAP*, 90, 5, 401, 1981.
244. Kochba, J. and Spiegel-Roy, P., Cell and tissue culture for breeding and development studies of *Citrus*, *HTCSA4*, 12, 2, 110, 1977.
245. Bonga, J. M., Clonal propagation of mature trees: problems and possible solutions, *Proc. Cell Tissue Culture Forestry*, 1, 249, 1987.
246. Karpechenko, G. D., Experimental polyploidy and haploidy, *Teor. Osn. Selek. Rast.*, 1, 397, 1935.
247. Takhtajan, A., Outline of the classification of flowering plants (Magnoliophyta), *BOREA4*, 46, 3, 226, 1980.
248. Naumova, T. N., Nucellar and integumentary embryony in angiosperms, *ASBNA2*, 50, 1, 213, 1981.
249. Randell, B. R., Adaptations in the genetic system of Australian arid zone *Cassia* species (Leguminosae, Caesalpinioideae), *AJBTAP*, 18, 1, 77, 1970.
250. Vasilyevskaja, V. K. and Borisovskaja, G. M., Life forms and its evolutionary reorganization in the family Buxaceae Dum., in *Life Forms: Structure, Spectrum, Evolution*, Nauka, Moscow, 1981, 90.
251. Tiagi, V. D., Cactacea, *Bull. Indian Natl. Sci. Acad.*, 41, 29, 1970.
252. Mauritzon, J., Über die Embryologie der Familie Rutaceae, *SBOTAS*, 29, 2, 319, 1935.
253. Fish, F. and Waterman, P. G., Chemosistematics in the Rutaceae. II. The chemosistematics of the *Zanthoxylum/Fagara* complex, *TAXNAP*, 22, 2–3, 177, 1973.
254. Waterman, P. G., Alkaloids of the Rutaceae: their distribution and systematic significance, *Biochem. Syst. Ecol.*, 3, 3, 149, 1975.
255. Naumova, T. N., Embryology of representatives of the family Celastraceae. The significance of integumental embryony in the evolution of genus *Euonymus*, in *Actualnye Voprosy Embryologii Pokrytosemennykh Rastenii*, Nauka, Leningrad, 1979, 46.
256. Leonova, T. G., *Euonymus Native to the USSR and Neighbouring Countries*, Nauka, Leningrad, 1974, 132.
257. Naumova, T. N., Family Celastraceae, in *Comparative Embryology of Flowering Plants (Davidiaceae-Asteraceae)*, Nauka, Leningrad, 1987, 49.

258. Brizicky, G. K., Polyembryony in *Euonymus* (Celastraceae), *JAARAG*, 45, 2, 206, 1964.
259. Brizicky, G. K., The genera of Celastrales in the Southeastern United States, *JAARAG*, 45, 2, 251, 1964.
260. Braun, A., *Über Polyembryonie und Keimung von Caelebogyne. Ein Nachtrag zur Abhandlung über Parthenogenesis bei Pflanzen*, Berlin, 1860, 263.
261. Petit-Thoars, A., Observations sur la germination de l'*Allim fragrans* et de quelques autres plantes dont les graines renferment plusieurs embryons distincts, *Nouv. Bull. Soc. Philom. (Paris)*, 12, 128, 1807.
262. Takhtajan, A., *The System of Magnoliophyta*, Nauka, Leningrad, 1987, 439.
263. Swamy, B. G. L., Embryological studies in the Orchidaceae, *AMNAAF*, 41, 202, 1949.
264. Mandrik, V. Yu., The forms of apomixis in the representatives of the family Rosaceae, *BYGBAA*, 116, 86, 1980.
265. Kholchlov, S. S., Asexual seed plants. Historical approach and evolutionary prospects, *USGCAA*, 16, 1, 3, 1946.
266. Rubtzova, Z. M., *Evolutionary Significance of Apomixis*, Nauka, Leningrad, 1989, 153.
267. Zavadsky, K. M., The problem of the species in apomictic plants, in *The Problems of Apomixis in Plants*, Thesis, Izd. Sarat. University, Saratov, 1966, 17.
268. Levina, R. E., *Reproductive Biology of Seed Plants*, Nauka, Moscow, 1981, 95.
269. Abeln, Y. S., Wilms, H. J., and van Wijk, A. J. P., Initiation of apomictic seed production in Kentucky bluegrass, *Poa pratensis* L., in *Proc. 8th Int. Symp.on Sexual Reproduction in Seed Plants, Ferns and Mosses*, Pudoc, Wageningen, 1984, 160.
270. Naumova, T. N., Apogamety in *Trillium camschatcense*: ultrastructural aspects, *ANL*, 3, 16, 1991.
271. Naumova, T. N., Apomixis and amphimixis in angiosperms; classification, *ANL*, 2, 33, 1990.
272. Crane, C. F., A proposed classification of apomictic mechanisms in flowering plants, *ANL*, 1, 11, 1989.
273. Shoda, S. P. and Bhanwra Ravinder, K., Apomixis in *Capillipedium huegelii* (Hack.) Stapf. (Gramineae), *PIBSBB*, 46, 4, 572, 1980.
274. Solntzeva, M. P., Disturbances in the process of fertilization in angiosperms under hemigamy (hemigamy and its manifestation in plants), in *Fertilization in Higher Plants*, Amsterdam, 1974, 311.
275. Koscinska-Pajak, M., Cytological patterns of the embryo structure in hemigamic plants, *ANL*, 3, 14, 1991.
276. Sen, P. K. and Mallik, P. C., The embryo of the Indian mangoes (*Mangifera indica* L.), *IJASA3*, 10, 750, 1940.
277. Gurgel, J. J. A., Polyembryionia e embryiogenia adventica em *Citrus, Mangifera* e *Eugenia*, *Rev. Lit. Duser.*, 3, 443, 1952.
278. Sobrinho, S. J. and Gurgel, J. T. A., Poliembrionia e embrionia adventicia em *Citrus, Mangifera* e *Murtaceae frutiferas, DUSEA5*, 4, 421, 1953.
279. Alexandrovsky, E. S. and Naumova, T. N., Family Anacardiaceae, in *Comparative Embryology of Flowering Plants (Brunelliaceae-Tremandraceae)*, Nauka, Leningrad, 1985, 166.
280. Hakansson, A., Studien über die Entwicklungsgeschichte der Umbelliferen, *LUARAS*, 2, 18, 7, 1, 1923.
281. Sufeldner, G., Die Polyembryonie bei *Cynanchum vincetoxicum* (L.) Pers., *Acad. Wiss. Wien. Math.-naturwiss. Kl.*, 1, 121, 273, 1912.
282. Naumova, T. N., Family Asclepiaceae, in *Comparative Embryology of Flowering Plants (Davidiaceae-Asteraceae)*, Nauka, Leningrad, 1987, 132.
283. Rudenko, K. Yu., Apomixis in some high-mountain plants of Ukraine Carpathians, *UKBZAW*, 18, 6, 24, 1961.
284. Maheshwari, D. H. and Pullaiah, T., Embryological investigations in the Melampodinae. I. *Melampodium divaricatum*, *PHYMAW*, 26, 1, 77, 1976.
285. Maheshwari, D. H. and Pullaiah, T., Embryology of Niger and Sanflower, in *Proc. Indo-Soviet Symposium on Embryology of Crop Plants*, Delhi, 1976, 21.

286. Korchagina, I. A., Family Betulaceae, in *Comparative Embryology of Flowering Plants (Winteraceae-Juglandaceae)*, Nauka, Leningrad, 1981, 201.
287. Robins, A., Note preliminare sur la polyembryonie dans *Bombax kimuenzae* de Wild., *JBNBA6*, 29, 1, 23, 1959.
288. Ghosh, R. B., Studies on the embryology of Burseraceae — embryogeny of *Garuga pinnate* Roxb., *BRORAF*, 39, 1–2, 17, 1970.
289. Orr, W., Polyembryony in *Sarcococca ruscifolia* Stapf., *NRYBAD*, 14, 21, 1923.
290. Naumova, T. N., Ovule, megasporogenesis, micro- and megagametogenesis of *Sarcococca humilis* Hort. (Buxaceae), *BOTZA9*, 64, 5, 635, 1979.
291. Naumova, T. N., Family Buxaceae, in *Comparative Embryology of Flowering Plants (Winteraceae-Juglandaceae)*, Nauka, Leningrad, 1981, 182.
292. Peter, J., Zur Entwicklungsgeschichte einiger Calycanthaceen, *BEPFAT*, 14, 59, 1920.
293. Schurhoff, P. N., Zur Apogamie von *Calycanthus*, *Flora*, 116, 73, 1923.
294. Mauritzon, J., Die Embryologie einiger Capparidaceen sowie von *Tovaria pendula*, *Arkiv. Bot.*, 26A, 15, 1, 1934.
295. Grebel, Dr., Über die Samen des *Evonymus latifolius*, *Flora*, 21, 321, 1920.
296. Jager, G. F., *Uber die Missbildungen der Gewächse*, Stuttgart, 1914, 40.
297. Bally, W., Basel, Zwei Fälle von Polyembryonie und Parthenokarpie, *VSNGAY*, 98, 169, 1916.
298. Adatia, R. D. and Gavde, S. G., Embryology of the Celastraceae, in *Plant Embryology*, New Delhi, 1962, 1.
299. Zaikovskaja, N. B., Peretjatjko, N. A., Shiryaeva, E. I., and Jarmoluk, G. I., Apomixis in the sugar beet forms with genic sterility of pollen, in *Apomixis and Cytoembryology of Plants*, Izd. Sarat. University, Saratov, 1978, 39.
300. Shiryaeva, E. I. and Yarmoluk, G. I., Adventive embryony and polyembryony of *Beta vulgaris* L., in *Thesis of VIII Vsesojusnyi Simposium po embriologii rastenii*, FAN, Tashkent, 1983, 152.
301. Sprecher, M. A., Etude sur la semence et la germination du *Garcinia mangostana* L., *BGBOA6*, 31, 513, 611 1919.
302. Mauritzon, J., Contributions to the embryology of the orders *Rosales* and *Myrtales*, *LUARAS*, 35, 2, 1, 1939.
303. Naumova, T. N., Family Combretaceae, in *Comparative Embryology of Flowering Plants (Brunelliaceae-Tremandraceae)*, Nauka, Leningrad, 1985, 101.
304. Yong, K. and Kaur, A., A cytotaxonomic view of the Dipterocarpaceae with some comments on polyploidy and apomixis, *Mem. Mus. Nat. Hist. Natur. Ser. Bot.*, 26, 41, 1979.
305. Ventura, M., Sulla poliembrionia di *Mallotus japonicus* Muell., *ANBOA4*, 20, 568, 1934.
306. Cesca, G., Ricerche cariologiche ed embriologiche sulle Euphorbiaceae. I. Su alcuni biotipi di *Euphorbia dulcis* L. della toscana, *CARYAB*, 14, 79, 161.
307. Symon, D. E., Polyembryony in *Cassia*, *NATUAS*, 177, 4500, 191, 1956.
308. Kobasnidze, E. Ya., The phenomenon of polyembryony in Grossulariaceae, in *Sbornik Trudov Molodykh Nauchnukh Rabotnikov Instituta Botaniki Akademii Nauk Gruzinskoi SSR*, Vol. 7, 1976, 70.
309. Kobasnidze, E. Ya., Embryology of Grossulariaceae in the High-Altitude Conditions of Georgia, Ph. D. dissertation, Tbilisi, 1983.
310. Subba Rao, A. M., Studies in the Malpighiaceae. I. Embryo sac development and embryology in the genera *Hiptage*, *Banisteria* and *Stigmatophyllum*, *JIBSAC*, 18, 145, 1940.
311. Subha Rao, A. M., Studies in the Malpighiaceae. II. Structure and development of the ovules and embryo sacs of *Malpighia coccifera* Linn. and *Tristellatela australis*, *PIIBAS*, 7, 393, 1941.
312. Singh, B., Studies in the family Malpighiaceae. I. Morphology of *Thryallis glauca*, *HTCAAI*, 3, 1, 1959.
313. Naumova, T. N., Family Malpighiaceae, in *Comparative Embryology of Flowering Plants (Brunelliaceae-Tremandraceae)*, Nauka, Leningrad, 1985, 222.

314. Prakash, N., Lim, L. L., and Menurung, R., Embryology of Duku and Langsat variation of *Lansium domesticum*, *PHYMAW*, 27, 1, 50, 1977.

315. Zamotailov, S. S., Embryology of Ficus in different variants of pollination, *IANBAM*, 2, 103, 1955.

316. Narayanan, C. K., Polyembryony in *Streblus taxoides* (Heyne) Kudz., *JIBSAC*, 47, 3–4, 354, 1969.

317. Zdruikovskaya-Rikhter, A. I., Apomixis in *Ficus afganistanica*, in *Apomixis and Cytoembryology of Plants*, Izd. Sarat. University, Saratov, 1978, IV, 45.

318. Kume, N., Morphologische-Physiologische Üntersuchungen über die Entwicklung von *Ardisia*, *Contrib. Biol. Lab. Kyoto Univ.*, 8, 1, 1959.

319. Traub, H. P., Polyembryony in *Myrciaria cauliflora*, *BOGAA5*, 101, 1, 233, 1939.

320. Johansen, D. A., Studies on the morphology of the Onagraceae. II. Embryonal manifestations of fasciation in *Clarkia elegans*, *BOGAA5*, 90, 75, 1930.

321. Roy, S. K., Embryology of *Eugenia fruticosa*, *PAIBA6*, 31, 1, 80, 1961.

322. Polunina, N. N., Biology and embryology of *Callistemon lanceolatus*, *BOTZA6*, 43, 8, 1169, 1958.

323. Narayanaswami, S. and Roy, S. K., Embryo sac development and polyembryony in *Syzygium cumini* (Linn.) Skeels., *BONOAE*, 3, 113, 273, 1960.

324. Mistra, R. S., Polyembryony in *Syzigium cumini*, *IJFIAW*, 5, 1, 69, 1982.

325. Francini, B., Fenomeni di aposporia somatica, di aposporia goniale e di embrionia avventzia in *Ochna multiflora*, *C.R. Acad. Lincei (Roma)*, 7, 92, 1928.

326. Pauze, F., Développement Floral et Embryologie d'*Ochna atropurpurea*, DC, doctoral thesis, McGill University, Montreal, 1972.

327. Naumova, T. N., Family Ochnaceae, in *Comparative Embryology of Flowering Plants (Phytolaccaceae-Thymelaeaceae)*, Nauka, Leningrad, 1983, 78.

328. Sinoto Vosito, On the nuclear divisions and partial sterility of *Oenothera lamarekiana* Ser., *BOMZA8*, 36, 92, 1922.

329. Michaelis, P., Zur Cytologie und Embryoentwicklung von *Epilobium*, *BEDBAP*, 43, 61, 1925.

330. Beth, K., Untersuchungen über die Auflösung von Adventivembryonie durch Wundreiz, *PLANAB*, 28, 296, 1938.

331. Zhgonti, L. P., Apomixis in genus *Paeonia*, *Vest. Gruz. Bot. Sada Acad. Nauk Gruz. SSR*, 7, 113, 1978.

332. Hafliger, E., Zytologisch-embryologische Untersuchungen pseudogamer Ranunculen der Auricomus Gruppe, *Ber. Schweiz. Bot. Ges.*, 53, 317, 1943.

333. Sokolovskaja, T. B., Family Ranunculaceae, in *Comparative Embryology of Flowering Plants (Winteraceae-Juglandaceae)*, Nauka, Leningrad, 1981, 130.

334. Souèges, R., Polyembryonie chez le *Potentilla reptans* L., *BSBFAN*, 82, 381, 1935.

335. Lebègue, A., Polyembryonie chez le *Potentilla aurea* L., *BSBFAN*, 97, 7–9, 199, 1950.

336. Cherneki, I. M., Apomixis in *Malus* of Zacarpathija, in *Apomixis and Cytoembryology of plants*, Izd. Sarat. University, Saratov, 1978, 4, 125.

337. Chakravarty, R. S., Nucellar polyembryony in Rutaceae, *CUSCAM*, 5, 202, 1936.

338. Torres, J. P., Polyembryony in *Citrus* and study of hybrid seedlings, *PHJAAN*, 7, 37, 1936.

339. Bacchi, O., Cytological observations in *Citrus*. III. Megasporogenesis, fertilization and polyembryony, *BOGAA5*, 105, 2, 221, 1943.

340. Rutishauser, A., *Embryologie und Fortpflanzungsbiologie der Angiospermen*, Springer-Verlag, Vienna, 1964, 163.

341. Mamporia, F. D., Cytoembryological investigations of *Poncirus trifoliata*, *TGSIARI*, 19, 39, 1943.

342. Moreira, S., Gurgel, I. T. A., and de Arruda, L. F., Polyembryony in *Citrus*, *BRGTAF*, 7, 69, 1947.

343. Furusato, K., Studies on polyembryony in *Citrus*, *Ann. Rep. Natl. Genet.*, 4, 56, 1951.

344. Banerji, J., Morphological and cytological studies on *Citrus grandis* Osbeck, *PHYMAW*, 4, 3–4, 390, 1954.

345. Minessy, F. A. and Higazy, M. K., Polyembryony in different *Citrus* varieties in Egypt, *AAGRAF*, 5, 89, 1957.
346. Kapanadze, I. S., On the question of polyembryony in *Citrus, SAKNAH*, 21, 2, 171, 1958.
347. Kapanadze, I. S., The phenomenon of multi embryony in Rutaceae, *SAKNAH*, 35, 2, 447, 1966.
348. Kapanadze, I. S., The development of nucellus and nucellar embryos in Satsuma orange, *SUKUA8*, 1, 77, 1983.
349. Hodgson, H. W., Horticultural applications of polyembryony in *Citrus, IJHOAQ*, 18, 4, 245, 1961.
350. Motial, V. S., Polyembryony studies in some of the limes, lemons and other citrus rootstock varieties, *SCINAL*, 29, 9, 1963.
351. Tzinger, N. V., Poddubnaja-Arnoldi, V. A., Petrovskaja, T. P., and Polunina, N. N., On the question of the apomixis causes (hystochemical investigation of generative organs of apomictic representatives of *Taraxacum* and *Citrus*), *TMPIAK*, 13, 201, 1965.
352. Tzinger, N. V.and Petrovskaja-Baranova, T. P., The physiological bases of apomixis in plants, *DASBAQ*, 172, 3, 737, 1967.
353. Sholokhova, V. A., On the question of polyembryony of the genus *Citrus*, in *Apomixis and Selection*, Kolos, Moscow, 1970, 220.
354. Iona, R.and Goren, R., Processing stripping film autoradiographs of *Citrus* buds. Avoidance of entrapped air between tissues and development embryo, *STTEAW*, 46, 3, 156, 1971.
355. Ghosh, R. B., Studies on embryogenesis of the Indian taxa of Rutaceae – a review, *AEECAM*, 13, 1–2, 201, 1975.
356. Bruck, D. K. and Walker, D. B., Cell determination during embryogenesis in *Citrus jambohiri*. I. Ontogeny of the epidermis, *BOGAA5*, 146, 2, 188, 1985.
357. Starrantino, A., Spino, P., and Russo, F., Nucellar embryogenesis and *in vitro* plant development from the nucellus of some species of *Citrus, J. Bot. Ital.*, 112, 1–2, 41, 1978.
358. Starrantino, A. and Reforgiato, R. G., Embriogenesi nucellare in ovuli non sviluppati di alcune cultivar di arancio del gruppo navel, *J. Bot. Ital.*, 117, 3–4, 107, 1983.
359. Ikeno, S., On hybridization of some species of *Salix, ANBOA4*, 36, 75, 1922.
360. Banerji, J. and Bhaduri, P. N., Polyembryony in Solanaceae, *CUSCAM*, 1, 10, 310, 1939.
361. Glutchenko, G. I., Early manifestation of adventive embryony in *Scopolia carniolica*, in *Apomixis in Animals and Plants*, Nauka, Novosibirsk, 1968, 10.
362. Gentry, H. S., Apomixis in black pepper and jojoba, *JOHEAS*, 46, 1, 8, 1955.
363. Gentry, H. S., The natural history of jojoba (*Simmondsia chinensis*) and its cultural aspects, *J. Econom. Bot.*, 12, 3, 261, 1958.
364. Davis, G. L., *Systematic Embryology of the Angiosperms*, New York, 1966, 505.
365. Naumova, T. N., Family Thymelaeaceae, in *Comparative Embryology of Flowering Plants (Phytolaccaceae-Thymelaeaceae)*, Nauka, Leningrad, 1983, 280.
366. Naumova, T. N., Are adventive embryony and gametophytic apomixis the different types of agamospermy?, *ANL*, 1, 55, 1989.
367. Phatak, V. G., Embryology of *Zygophyllum coccineum* L. and *Z. fabago* L., *KNWCAD*, 74, 4, 379, 1971.
368. Hakansson, A., Parthenogenesis in *Allium, BONOAE*, 1, 143, 1951.
369. Bambacioni-Mazetti, V., Sullo sviluppo dell'embrione in *Tulipa gesneriana* L., *ANBOA4*, 19, 145, 1931.
370. Suessenguth, K., Über die Pseudogamie bei *Zygopetalum mackayi Hook.*, *BEDBAP*, 41, 1, 16, 1923.
371. Seshagiriah, K. N., Development of the female gametophyte and embryo in *Spiranthes australis* (Lindley), *CUSCAM*, 1, 4, 102, 1932.
372. Seshagiriah, K. N., Morphological studies in Orchidaceae. I. *Zeuxine sulcata* Lindley, *JIBSAC*, 20, 357, 1941.
373. Swamy, B. G. L., Agamospermy in *Spiranthes cernua* (Orchidaceae), *LLOYA2*, 11, 3, 149, 1942.

374. Dieter, I. R., Sobre a reproducao em *Maxillaria brasiliensis* Brieg., *RBBIAL*, 37, 2, 267, 1977.
375. Vij, S. P., Sharma, M., and Shekhar, N., Embryological studies in Orchidaceae. II. *Zeuxine strateumatica* complex, *PHYMAW*, 32, 2–3, 257, 1982.
376. Longping Yuan, Yuanching Li, Hongde Deng, Progress of studies twin seedlings, *ANL*, 2, 42, 1990.
377. Jeffrey, E. C., and Haertl, E. J., Apomixis in *Trillium*, *CELLA*, 48, 79, 1939.
378. Swamy, B. G. L., On the post-fertilization development of *Trillium undulatum*, *CELLA*, 52, 7, 1947.
379. Naumova, T. N., Family Trilliaceae, in *Comparative Embryology of Flowering Plants (Butomaceae-Lemnaceae)*, Nauka, Leningrad, 1990, 151.
380. Dzevaltovskij, A. K., Family Cucurbitaceae, in *Comparative Embryology of Flowering Plants (Phytolaccaceae-Thymelaeaceae)*, Nauka, Leningrad, 1983, 127.
381. Pieringer, A. P. and Edwards, G. J., Identification of nucellar and zygotic *Citrus* seedlings by infrared spectroscopy, *PASHA6*, 86, 226, 1965.

ABBREVIATIONS

AE   Aposporic embryo sac
AN   Antipodal nuclei
CC   Central cell
E   Egg cell
EA   Egg apparatus
EC   Initial cell of nucellar or integumentary embryo = embryocyte
EM   Embryoderm
ES   Embryo sac
I   Inner integument
IE   Integumentary embryo
LI   Lateral part of integument
ME   Megasporocyte
MC   Meristematic cells
MIT   Meristematic inner integument tissue
MN   Micropillar part of nucellus
MNT   Meristematic nucellus tissue
MT   Megaspore tetrad
EN   Endosperm
NU   Nucellus
NE   Nucellar embryo
PR   Parietal tissue
PN   Polar nuclei
PT   Pollen tube
S   Synergid
SA   Shoot apex
Z   Zygote
ZE   Zygotic embryo
Ch   Chromosome
CW   Cell wall
D   Dictyosome
ER   Endoplasmic reticulum
LB   Lipid body
M   Mitochondria
N   Nucleus
NC   Nucleolus
NM   Nuclear membrane
P   Plastid
PB   Protein body
PD   Plasmodesmata
R   Ribosomes
RER   Rough endoplasmic reticulum
SER   Smooth endoplasmic reticulum
V   Vacuole

Apomixis in Angiosperms

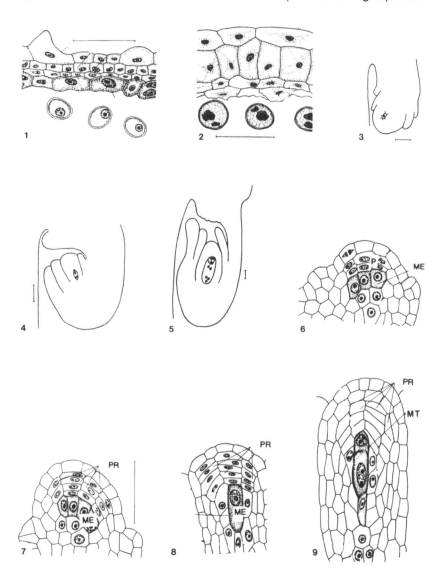

SCHEME I.   Reproductive structures of the genus *Sarcococca:* (1) pollen sac with unicellular pollen grains; (2) pollen sac with degenerating pollen grains; (3 to 5) ovule development; (6 and 7) ovule with three to four megasporocytes, parietal tissue is formed; (8) ovule with megasporocyte; (9) ovule with tetrad of megaspores, middle one is functional; (10 to 12) ovules with one, two, and four nuclei embryo sacs; (13) ovule with mature embryo sac, polar and antipodal nuclei are in central cell; (14 and 15) abnormalities of megagametogenesis; (16) two-cellular endosperm, egg apparatus degenerates; (17) cellular endosperm, egg apparatus degenerates; (18) contact of polar and antipodal nuclei; (19) nuclei of two-cellular endosperm; (20) nuclei of multicellular endosperm; (21 to 38) development of nucellar embryos; (39) nucellar embryo of mature seed. (Parts 1 to 31, 34, 35, and 37 to 39 — *S. humilis;* parts 32, 33, and 36 — *S. hookeriana.*)

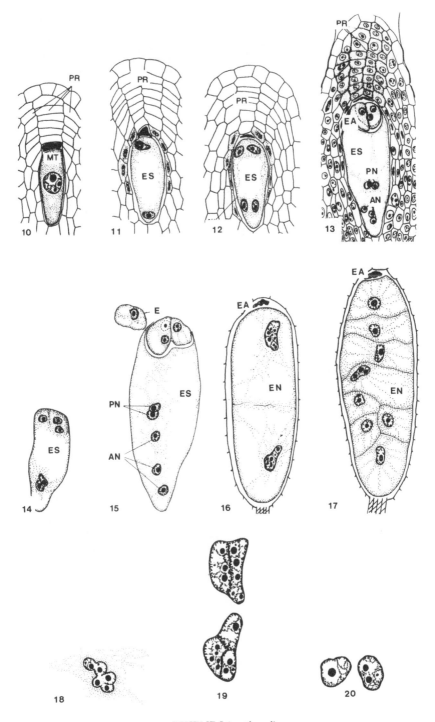

SCHEME I (continued)

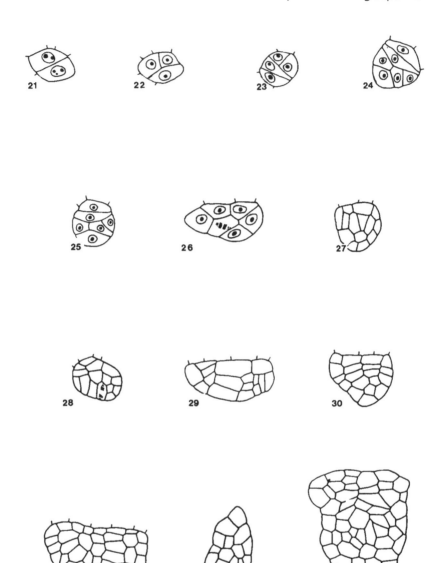

SCHEME I (continued)

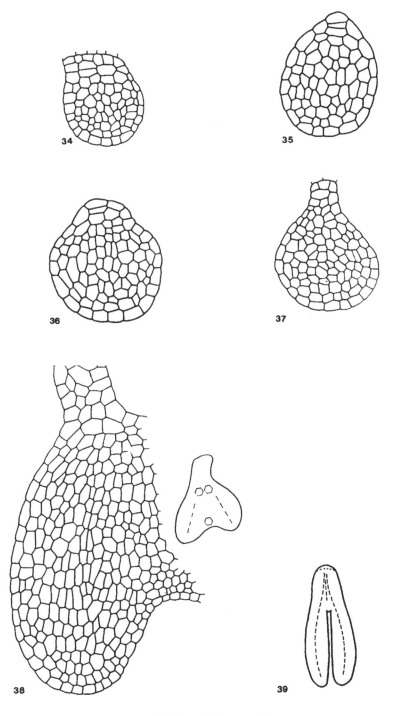

SCHEME I (continued)

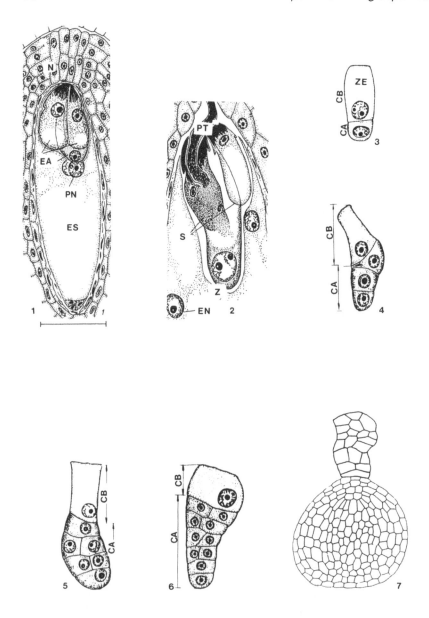

SCHEME II.   Reproductive structures of the genus *Opuntia:* (1) mature embryo sac; (2) embryo sac with zygote and nuclear endosperm, fragments of pollen tube and synergids are observed; (3 to 10) zygotic embryo development; (11) ovule with megasporocyte; (12) ovule with two-nuclei embryo sac; (13) ovule with mature embryo sac; (14 to 19) development of nucellar embryos; (20 to 24) successive stages of ovule development, blossom period (20), 7, 14, 21, and 35 days after pollination, respectively (21 to 24); (Parts 1 to 10 — *O. ficus-indica;* parts 11 to 24 — *O. elata.*)

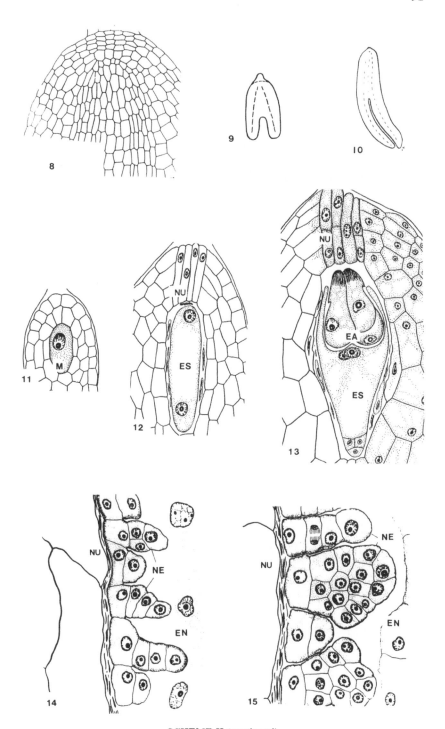

SCHEME II (continued)

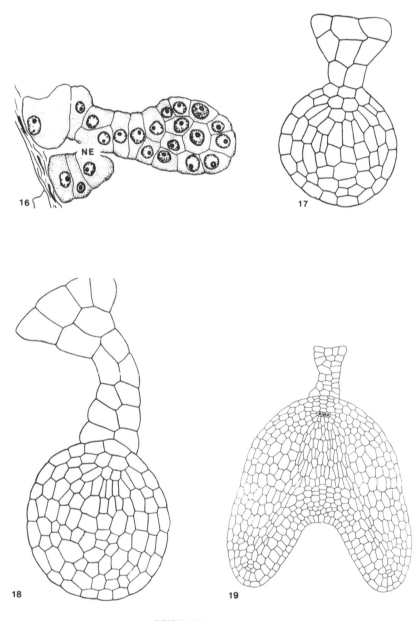

SCHEME II (continued)

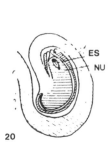

20

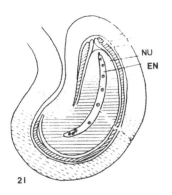

21

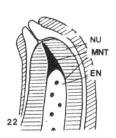

22

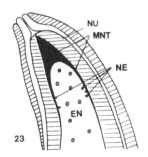

23

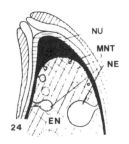

24

SCHEME II (continued)

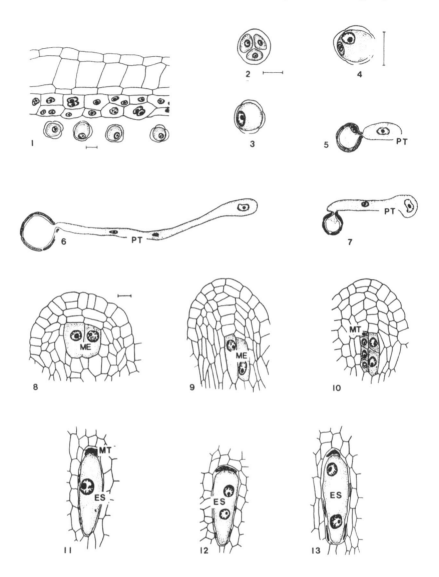

SCHEME III.   Reproductive structures of the genera *Citrus, Poncirus,* and *Fortunella:* (1) pollen sac with unicellular pollen grains; (2) tetrad of microspores; (3) unicellular pollen grain; (4) two-celled mature pollen grain; (5 to 7) germinating pollen grains; (8 and 9) ovules with megasporocytes; (10) ovule with tetrad and dyad of megaspores; (11 to 14) embryo sac development; (15) mature embryo sac; (16) embryocytes in nucellus, embryo sac is degenerating; (17) egg cell and nuclear endosperm; (18) zygote and central cell with polar nuclei; (19 to 22) micropiliar part of nucellus with zygote (21), embryocytes (19 and 21), nucellar (19, 20, and 21), and zygotic embryos (19 and 20); endosperm is nuclear in all cases; (23 to 33) development of zygotic embryos. (Parts 1 to 4 — *Fortunella margarita;* 5, 6, and 12 to 15 — *C. limon;* parts 7 and 21 — *C. reticulata;* parts 8 to 11, 23, 25 to 27, 30 to 33 — *C. limon* (Ponderosa); 24, 28, and 29 — *C. limon* (Meier); part 22 — *C. limon* (Novogrusinskii); parts 19 and 20 — *Poncirus trifoliata;* parts 16 to 18 — *C. limon* (Monakello).)

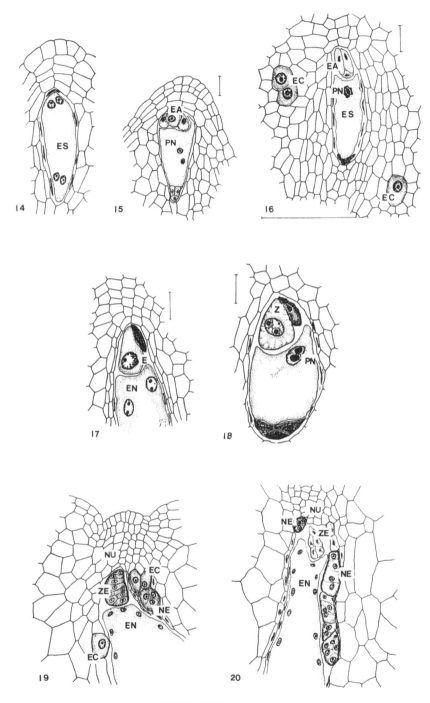

SCHEME III (continued)

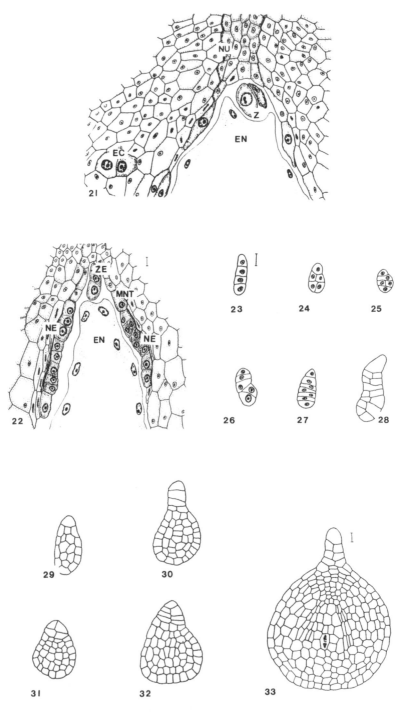

SCHEME III (continued)

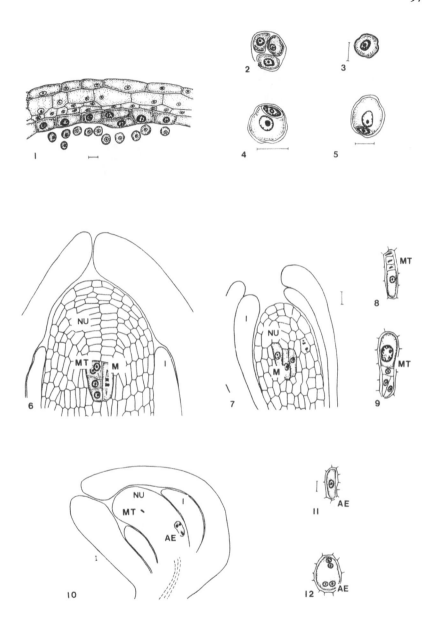

SCHEME IV. Reproductive structures of the genus *Zanthophylum:* (1) pollen sac with unicellular pollen grains; (2 to 5) pollen grains from tetrad-stage to two-celled stage; (6 and 7) micropillar part of ovule with tetrad of megaspores and megasporocyte (6) with three megasporocytes (7); (8 and 9) tetrad of megaspores with different position of functional megaspore; (10) ovule with degenerating tetrad of megaspores and four-nuclei aposporic embryo sac; (11 and 12) one- and four-nuclei aposporic embryo sacs; (13 and 14) micropillar part of nucellus; egg apparatus is degenerating; endosperm is nuclear. (Parts 1 to 4 — *Z. schinifolium;* parts 5, 6, and 8 to 12 — *Z. americanum;* part 7 — *Z. ailantifolium;* part 13 — *Z. simulans;* part 14 — *Z. spinifex*).

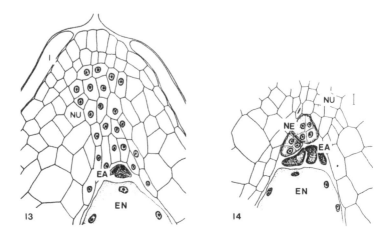

SCHEME IV (continued)

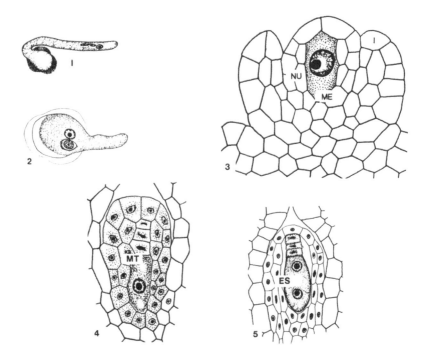

SCHEME V.   Reproductive structures of the genus *Euonymus:* (1 and 2) germinating pollen grains; (3) ovule with megasporocyte; (4) nucellus with tetrad of megaspores; the chalazal megaspore is functional; (5) nucellus with two-nuclei embryo sac; (6) four-nuclei embryo sac; (7) mature embryo sac; (8) embryo sac with developing endosperm; the egg apparatus is intact; (9) embryo sac with zygote and nuclear endosperm; (10 to 12) zygotic embryos and nuclear endosperm; (13 to 17) development of zygotic embryos; (18) mature seed; (19 to 37) development of integumentary embryos; (38 to 42) polyembryony in mature seeds. (Parts 1, 5 to 7, and 19 to 42 — *E. macroptera;* parts 2 and 8 — *E. planipes;* parts 3, 4, and 9 to 18 — *E. europaea.*)

99

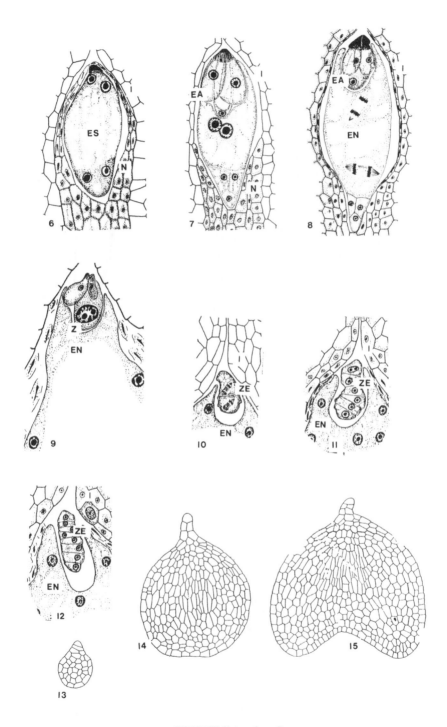

SCHEME V (continued)

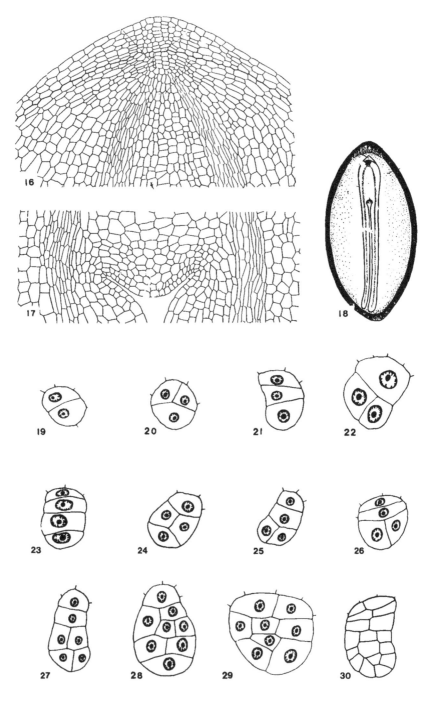

SCHEME V (continued)

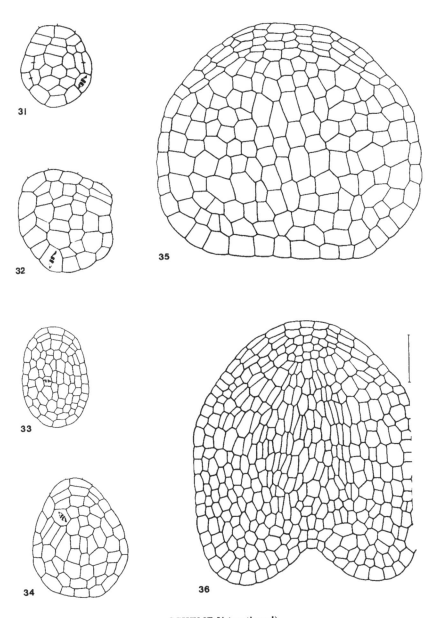

SCHEME V (continued)

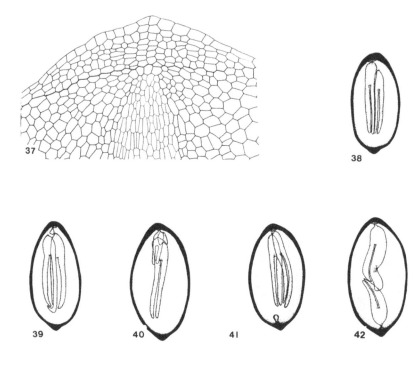

SCHEME V (continued)

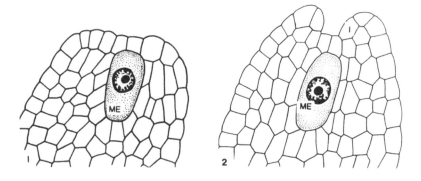

SCHEME VI. Reproductive structures of *Vincetoxicum officinale:* (1 and 2) nucellus with megasporocyte; (3) ovule with tetrad of megaspores; functional megaspores are chalazal and micropillar; (4) ovule with mature embryo sac; (5) micropillar part of integument with two integumentary embryos; endosperm is cellular.

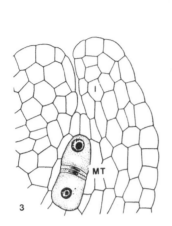

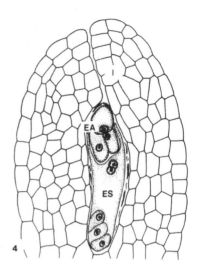

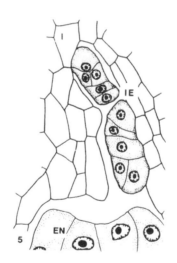

SCHEME VI (continued)

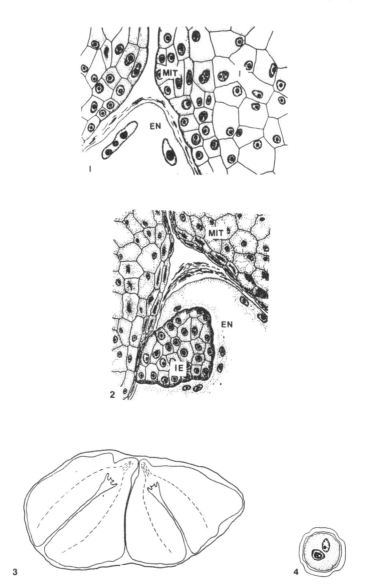

SCHEME VII.　　Reproductive structures of *Ochna multiflora:* (1) micropillar part of inner integument; endosperm is nuclear; egg apparatus degenerated; (2) fragment of inner integument with integumentary embryos; endosperm is nuclear; (3) mature seed with two embryos; (4) mature pollen grain.

105

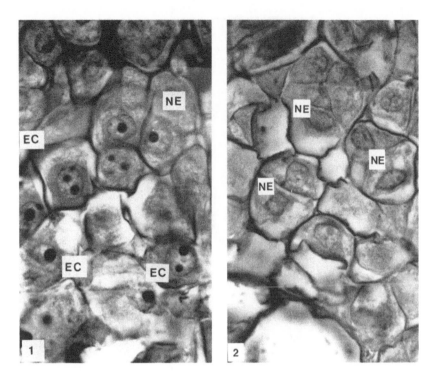

PLATE 1. Reproductive structures of *Sarcococca humilis:* (1) micropillar part of nucellus with embryocytes; shortly before division and with nucellar embryos; (2) micropillar part of nucellus with numerous nucellar embryos; their outer cell wall is thicker than inner ones; (3) multicellular nucellar embryos; (4) maturing seed with numerous nucellar embryos; (5) embryocyte; (6 to 10) fragments of embryocytes; nucleus with nucleolus; nuclear membrane possesses numerous pores (6); elongated mitochondria and oval plastids are located near the nucleus; polysomes are noticeable (7); cell wall with plasmodesmata, mitochondria with numerous cristae (8); cytoplasm with dictyosomes, numerous polysomes, and lipid bodies, vacuoles with inclusions (9); elongated plastids are near the nucleus (10); (11) embryocyte in prophase of mitosis; cell organelles are concentrated around the nucleus; vacuoles are on cell periphery; lipid bodies are numerous; (12) fragment of prophase cytoplasm with abundant plastids, mitochondria, and endoplasmic reticulum; (13) embryocyte in metaphase of mitosis; vacuoles of different sizes are numerous; they occupy most of the cell; cytoplasm with the cell organelles and chromosomes are in the cell center only; (14) fragment of cytoplasm in metaphase; mitochondria and plastids are changed morphologically; (15) two-celled nucellar embryo with thickened outer cell wall; while cell wall between embryo cells is thinner; (16) fragment of two-celled nucellar embryo cytoplasm; cell wall between embryo cells is developing; (17 and 18) fragment of three-celled nucellar embryo cytoplasm; outer cell wall is thickened; plasmodesmata are lacking; plastids and mitochondria are numerous; (19) fragment of micropillar part of nucellus with some developing nucellar embryos; the latter are adjacent to each other; their outer cell walls are thickened and lack plasmodesmata; (20) part of nucellus in the direction from micropyle to endosperm; numerous developing nucellar embryos are adjacent to each other; they show none of orientation toward the endosperm and no sequence in cell divisions; their outer cell walls are thick and lack plasmodesmata; (21 to 23) multicellular nucellar embryo; outer cell wall is thick, without plasmodesmata (21); fragments of embryo cytoplasm (22 and 23).

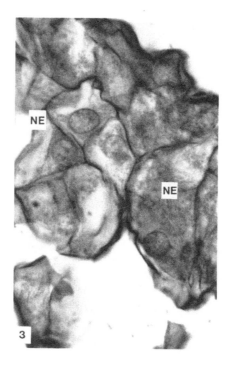

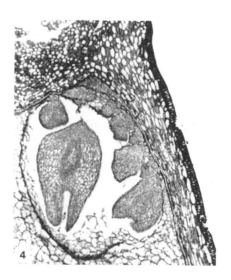

PLATE 1 (continued)

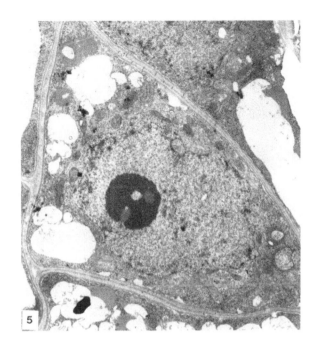

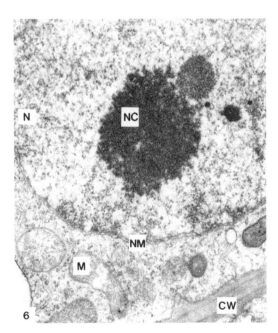

PLATE 1 (continued)

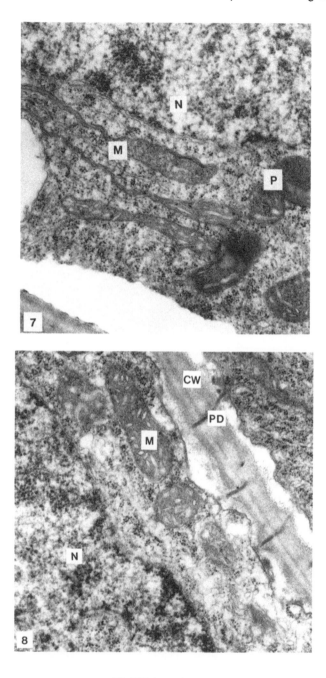

PLATE 1 (continued)

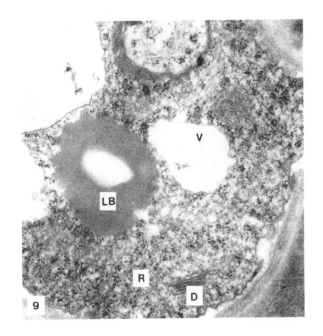

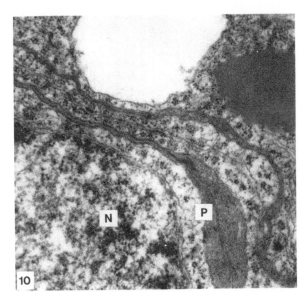

PLATE 1 (continued)

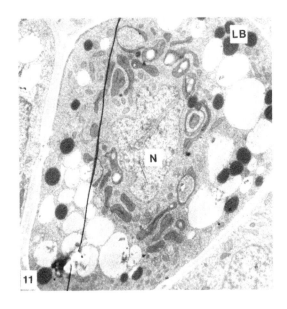

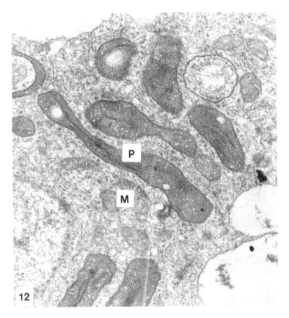

PLATE 1 (continued)

111

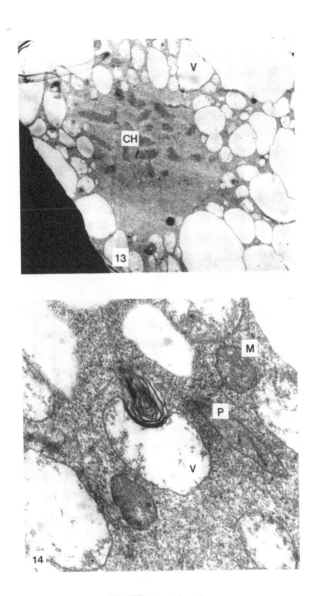

PLATE 1 (continued)

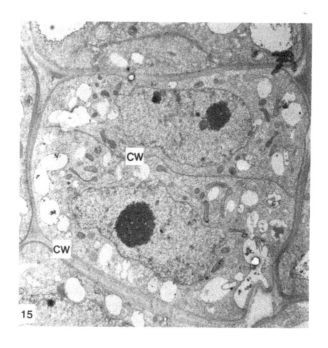

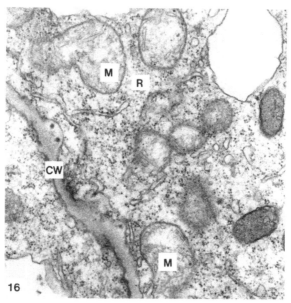

PLATE 1 (continued)

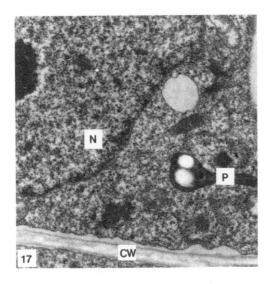

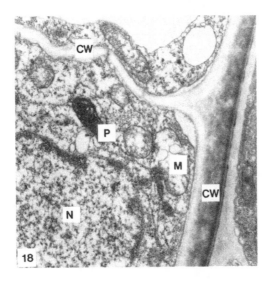

PLATE 1 (continued)

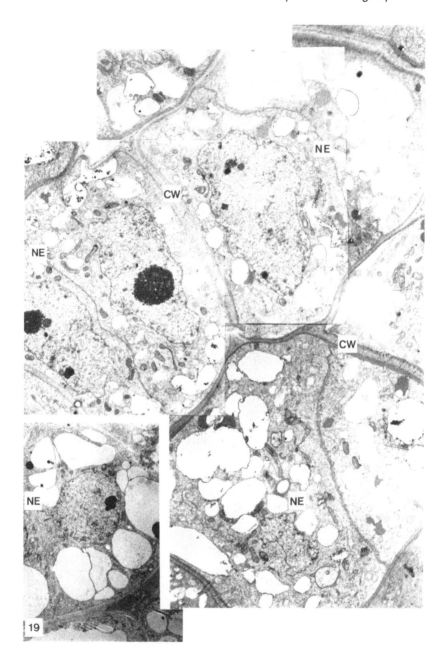

PLATE 1 (continued)

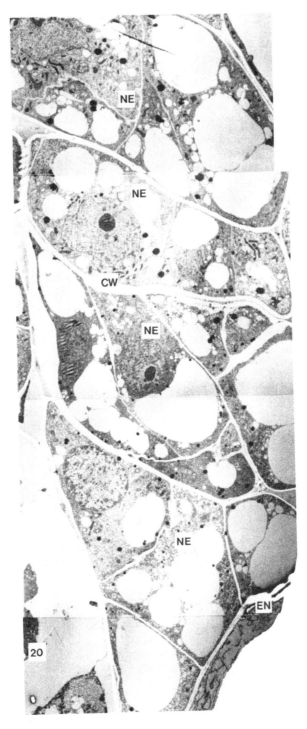

PLATE 1 (continued)

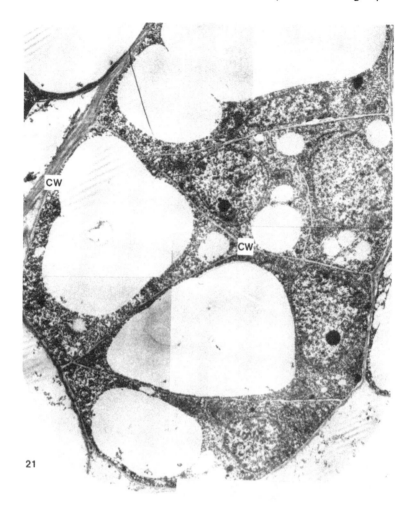

PLATE 1 (continued)

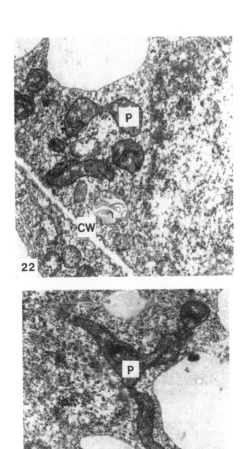

PLATE 1 (continued)

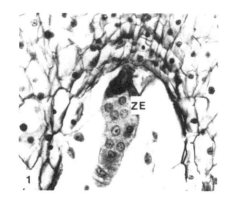

PLATE 2.   Reproductive structures of the genus *Opuntia:* (1 to 3) developing zygotic embryos; (4) micropillar part of nucellus with dividing dense cytoplasm cells; egg apparatus has degenerated; (5) part of the ovule with meristematic nucellar tissue adjacent to the nuclear endosperm; other nucellar cells are large and vacuolated; (6 and 7) fragment of seed with numerous developing nucellar embryos in lateral position. (Parts 1 to 3 — *O. ficus-indica;* parts 4 to 7 — *O. elata*).

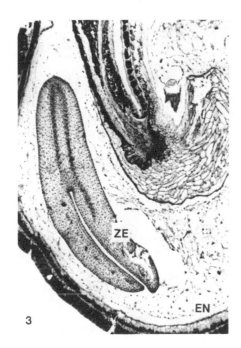

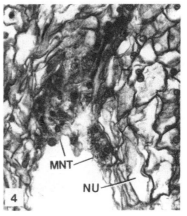

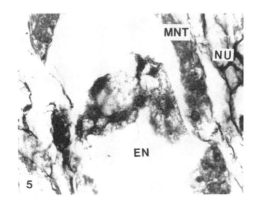

PLATE 2 (continued)

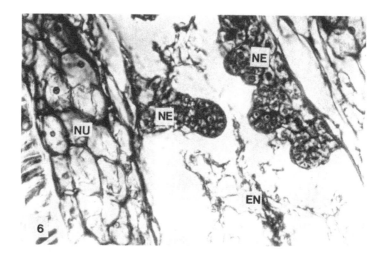

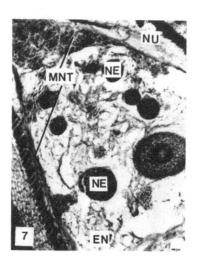

PLATE 2 (continued)

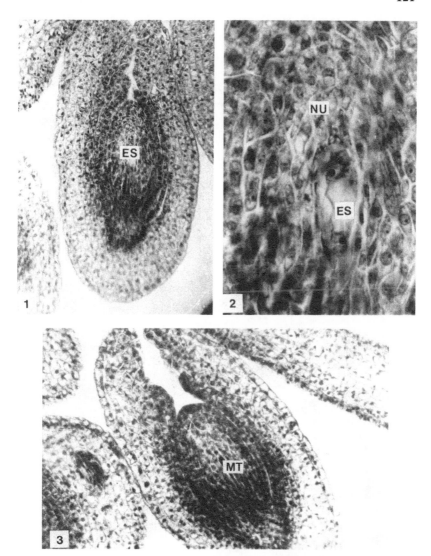

PLATE 3. Reproductive structures of the genera *Citrus, Poncirus,* and *Zanthoxylum:* (1 and 2) ovules with two-nuclei embryo sacs; (3) ovule with degenerating megaspore tetrad; (4) embryo sac with zygote and polar nuclei in central cell; remnants of pollen tube are close to zygote; (5) micropillar part of nucellus with dense cytoplasm cell nearest to the central cell; egg cell is in regular position; (6) embryocytes are in immediate vicinity of other nucellus cells (right) and endosperm (left); (7) embryo sac with endosperm and egg cell, which is very similar to the embryocytes; (8) ovule with nucellus which is of different cell patterns: dense cytoplasm cells are in micropillar part of nucellus, while its other cells are large and vacuolated; (9) fragment of nucellus with different cell pattern shown above; (10) two nucellar embryos are near nucellus cells (right) and endosperm (left); (11 and 12) developing seeds with some nucellar embryos showing different stages of development; (13) micropillar part of nucellus with developing nucellar embryos; endosperm is cellular; nucellus shows different cell patterns; (14) nucellar embryo and cellular endosperm. (Parts 1 to 3 — *C. limon* (Ponderosa), *C. limon* (Monakello); parts 5 to 12 — *P. trifoliata;* 13 and 14 — *Z. simulans.*)

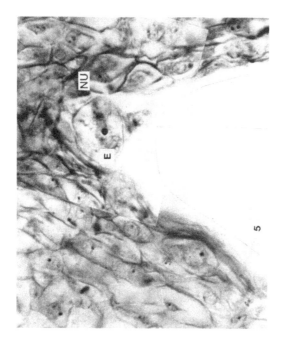

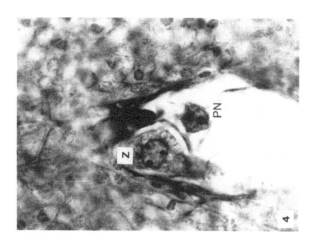

PLATE 3 (continued)

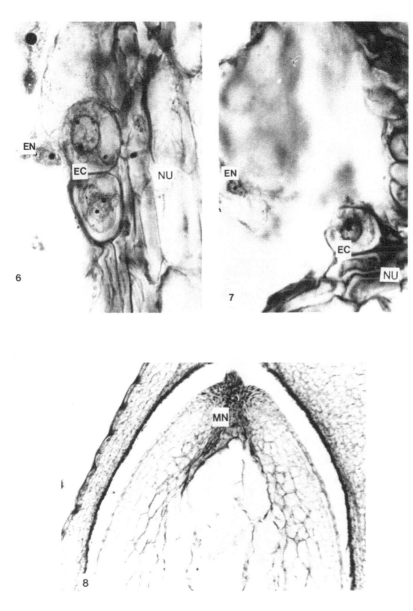

PLATE 3 (continued)

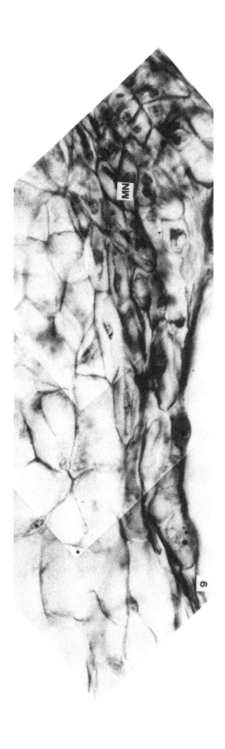

PLATE 3 (continued)

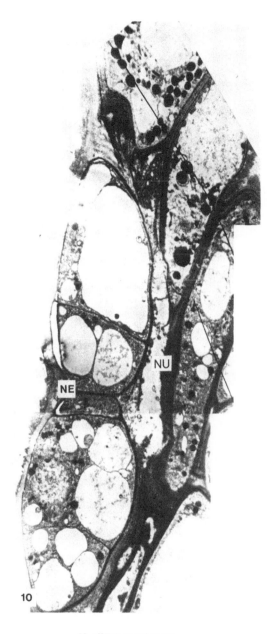

PLATE 3 (continued)

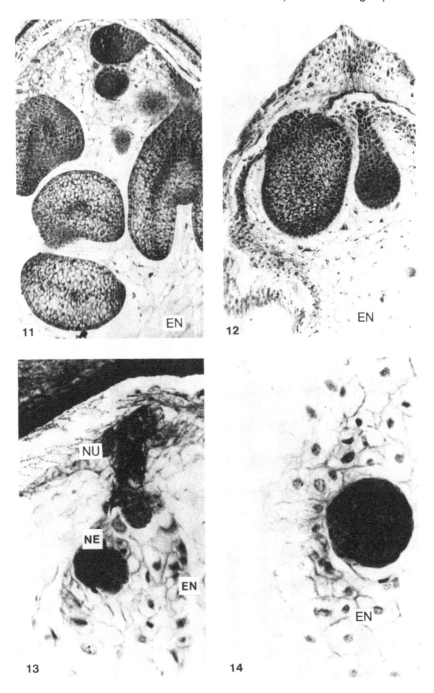

PLATE 3 (continued)

127

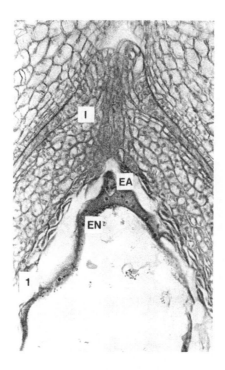

PLATE 4. Reproductive structures of *Euonymus macroptera:* (1) ovule after flowering; nucellus has degenerated; inner integument shows dense cytoplasm cells along micropyle; egg apparatus is degenerating; endosperm is nuclear; (2) micropillar part of inner integument; meristematic tissue is formed by dense cytoplasm cells; zygotic embryo and endosperm are present (stained by OsO₄); (3) micropillar part of inner integument with embryocytes and two-celled integumentary embryos; endosperm is nuclear; (4) fragment of inner integument with two-celled and multicelled integumentary embryos; endosperm is present; (5) micropillar part of ovule; inner integument is neighboring the cellular endosperm; two integumentary embryos are globular; remnants of pollen tube are visible; (6) embryocyte, fine structure; (7 to 9) fragments of embryocyte; part of nucleus; cytoplasm with numerous mitochondria and plastids; polysomes are abundant; vacuoles with inclusions; cell wall thickened; plasmodesmata are not observed (7); embryocyte cell wall with dark depositions from the endosperm side (8); cytoplasm with cup-like plastids, oval mitochondria, endoplasmic reticulum, and polysomes (9); (10) fragment of inner integument which is different pattern cells: lateral inner integument cells with protein storage are large; meristematic inner integument tissue is close to endosperm; embryocytes and integumentary embryos arising from this tissue cell; (11) micropillar part of seed with two integumentary embryos which are at different developmental stages; (12) maturing seed with three integumentary embryos, two are in the micropillar part and one is in the chalazal part of the same seed; (13) two integumentary embryos dissected from mature seed; (14) embryocytes of the chalazal part of inner integument; (15 and 16) embryocytes stained by procion dyes;

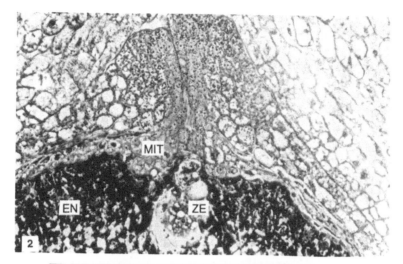

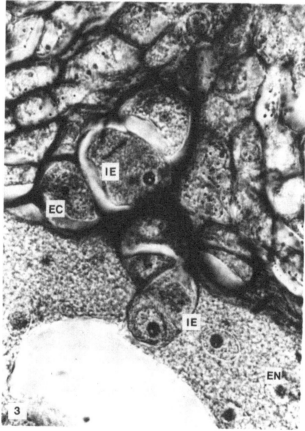

PLATE 4 (continued)

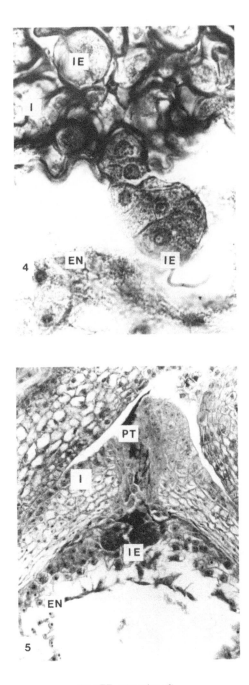

PLATE 4 (continued)

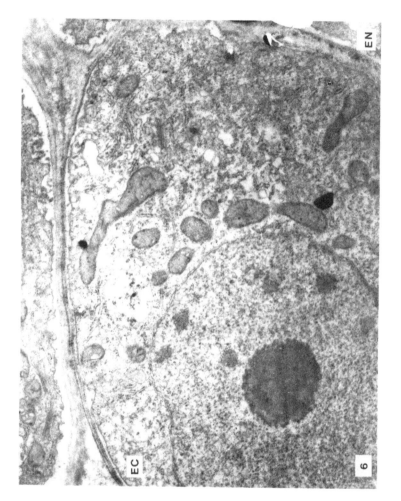

PLATE 4 (continued)

PLATE 4 (continued)

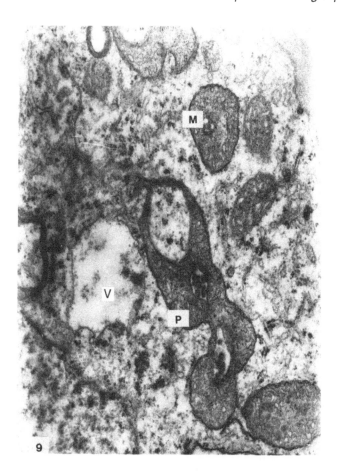

PLATE 4 (continued)

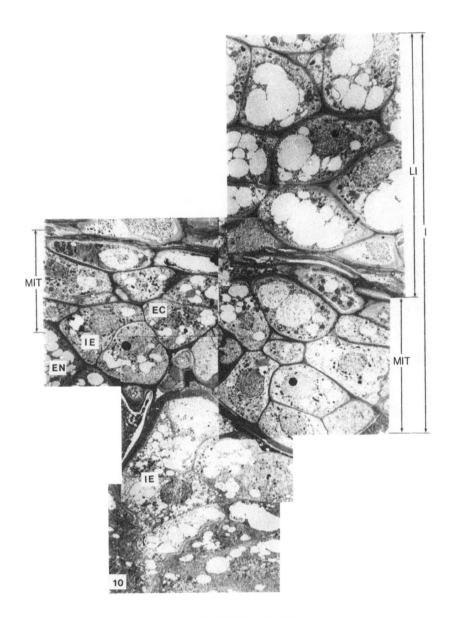

PLATE 4 (continued)

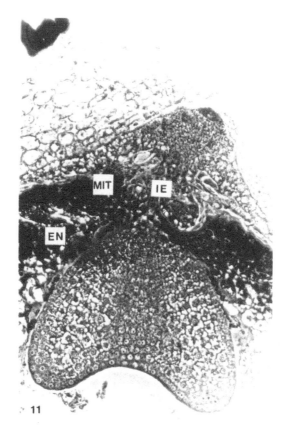

PLATE 4 (continued)

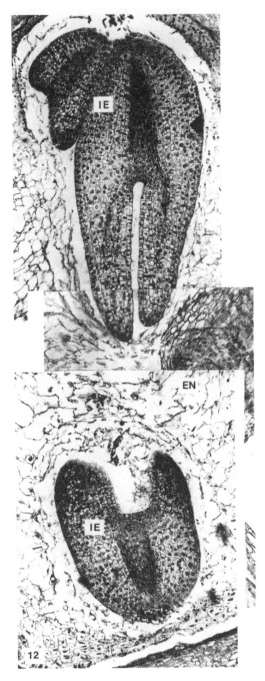

PLATE 4 (continued)

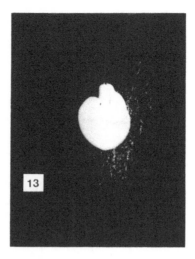

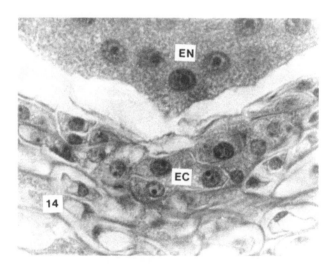

PLATE 4 (continued)

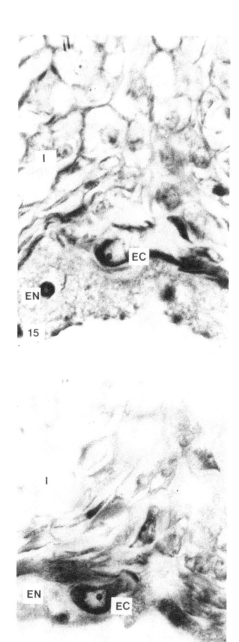

PLATE 4 (continued)

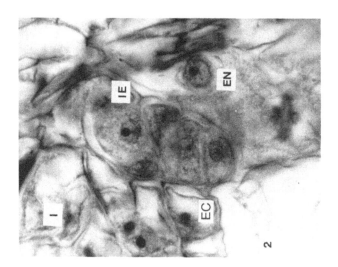

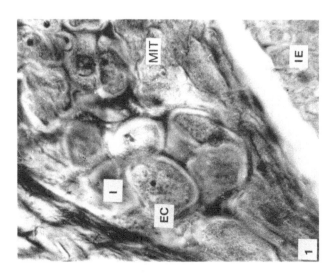

PLATE 5.   Reproductive structures of *Vincetoxicum officinale*: (1) inner integument with numerous embryocytes during seed maturation; (2) inner integument with integumentary embryo and embryocytes; endosperm is nuclear; (3) inner integument with some integumentary embryos; endosperm is nuclear; (4) developing seed with globular integumentary embryo; endosperm is nuclear; (5) two integumentary embryos dissected from mature seed.

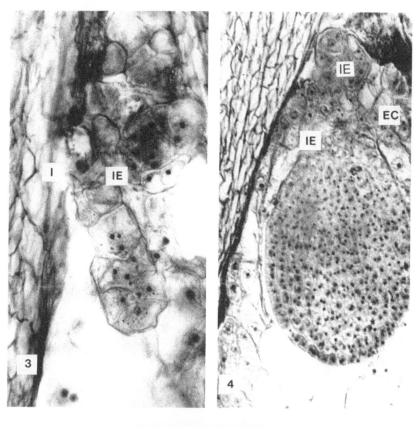

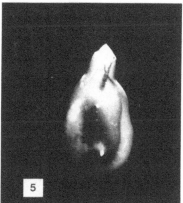

PLATE 5 (continued)

INDEX

A

Adventive embryo (nucellar, integumentary embryo), 27
  cell pattern, 27
  classification, 7
  development, 8, 40
  polarization, 46
  ultrastructure, 45
Adventive embryony (nucellar, integumentary embryony), 11, 57, 61
  in angiosperms, 43, 49
  in classification, 7, 62
  detecting method, 13
  in families, 49, 52
  in genera, 4, 38, 49, 52
  in species, 4, 49
Agamospermy, 3, 5, 10, 57, 58
Alternation of generations, 5
Amphimixis, 4, 23, 31, 44, 51, 52, 57
Androgenesis, 62
Aneuspory (diplospory), 11, 59
Angiosperms,
  general aspects, 1–2
  historical background, 3–12
Apogametophytic sporophyty (adventive embryony), 7
Apogamety, 5, 6, 11, 61, 63–70
Apogamy (diplospory), 5
Apomicts, diploid, 51
Apomixis, 23, 31, 44, 57
  autonomous, 5, 6
  classification, 4, 5, 7, 11
  embryology, 7
  forms, 4, 10, 59, 63–70
  gametophytic, 6
  induced, 5–6
  obligate, 52
  terminology, 4
  types, 9, 49, 59, 63–70
Apospory (somatic apospory, somatic evapospory, evapospory), 5–7, 11, 20, 44, 59
Apozygoty (parthenogenesis), 61
Archesporial cell (female), 16, 31

C

Callus, 45
Cell differentiation, 31
  during embryocyte development, 46–47

during embryoid (somatic embryo) development, 46
during nucellar, integumentary embryo development, 24–27
during zygotic embryo development, 37
Cell division, 34, 37
Cell division, unequal, 37
Cell polarity, 26
Cell wall, 26, 31, 57
Cytoplasm (protoplast), 31
  adventive embryos (nucellar, integumentary embryos), 46
  embryocytes (initial cells of nucellar, integumentary embryos), 31
  integumentary cells, 44
  nucellus cells, 30

D

Dictyosomes, 25, 30
Diplospory (apogamy, generative apospory, semiapospory, aneuspory, pseudoapospory), 5, 7, 11, 20, 58
Double fertilization, 21–22
Double fertilization, abnormalities, 22

E

Embryocyte (initial cells of nucellar, integumentary embryos), 2, 27, 35
  cell wall, 26, 27
  differentiation, 24, 27
  factors might promote the emergence, 32
  location, 40
  mitosis, 31
  polarity, 31, 40
  time, 33
  ultrastructure, 24, 35
Embryo development (embryogenesis), 9
  adventive (nucellar, integumentary), 37, 38, 40
  cell pattern, 36
  classifications, 40
  deviation from usual pattern, 19
  integumentary, 40
  nucellar, 35, 40
  sexual (zygotic), 37, 39
  somatic (embryoid), 46
  ultrastructural aspects, 35–36
Embryo-endosperm relation, 23
  in amphimicts, 23

in case of apomixis, 23
Embryogenesis (embryo development), 36, 40, 41
Embryoid (somatic embryo), 45
Embryony,
  list of species with nucellar and integumentary, 63–70
  occurrence of nucellar and integumentary, 49–62
    evolutionary significance of, 56
    within families, genera, and species of flowering plants, 51–56
    in system of flowering plants, 49–51
  structural and functional aspects of
    nucellar and integumentary, 13–47
    adventive embryos and embryoids, 45–47
    development of sexual and asexual embryos, 35–41
    double fertilization, 21–22
    endospermogenesis, 22–24
    initial cells of nucellar and integumentary embryos, 24–29
    materials and methods of research, 13–15
    megasporogenesis, megagametogenesis, and mature embryo sacs, 16–21
    microsporogenesis,
      microgametogenesis, and mature pollen grains, 15–16
    mitosis of embryocytes in *Sarcococca humilis*, 30–35
    polyembryony, 41–43
    theoretical grounds for nucellar and integumentary embryony, 43–45
Embryo sac (megagametogenesis), 17, 27
  abnormalities, 4, 9, 19
  antipodal cells, 10, 17
  egg cell, 17, 21
  mature, 19
  one- two- four-nuclei (coenocyte stage), 17, 34
  synergid, 21
  synergid embryo, 4, 9
Embryo sac (typology), 17
  aposporous, 63–70
  diplosporous, 63–70
  Polygonum-type, 17, 19
  tetrasporic, 18
  unreduced (bisporic), 18, 20
Endoplasmic reticulum, 25, 30, 57
Endosperm, 22
  apomictic, 22, 23
  autonomous development (without fertilization), 23

ploidy, 22
role in embryo development, 23
Evapospory (apospory), 59

F

Female gametophyte, 16
  angiosperms (embryo sac), 43, 45
  gymnosperms, 43, 45
Flowering plants, apomixis and amphimixis in seed reproduction of, 57–62

G

Gametogenesis, 15
Gametophytic apomixis, 6, 11, 57
Generative apospory, 6, 59

H

Hemigamy (semigamy), 11, 61
Heterophasic reproduction, 8, 9
Homophasic reproduction, 8, 9

I

Initial cell
  of aposporic embryo sac, 61
  of embryoid, 46
  of integumentary embryo (embryocyte), 24–29, 35
  of nucellar embryo (embryocyte), 24–29, 35
Integument, 34
Integumentary embryo, 34
  cell pattern, 35
  cell wall, 36
  development, 23
  polarization, 40
  time, 33
Integumentary embryony (adventive embryony), 63–70

L

Lipid bodies, 27, 36

M

Megasporangium (nucellus), 43, 57
Megaspore, 18, 19
  functional, 19
  position in tetrad, 17–19, 34
  tetrad (tetrad of megaspore), 18
  degeneration, 17

development, 18, 19
  ultrastructure, 19
Megasporocyte, 5, 16, 59
  meiosis, 30
  polarity, 31
Megasporogenesis, 30, 57
  archesporial cell (archesporium), 17
  parietal cells (parietal tissue), 16
  pattern of, 18
Meiosis, 15, 30, 31
  prophase I of, 19
  suppression of, 17
Microgametogenesis (pollen grains), 16
Microsporogenesis, 15, 16
Mitochondria, 25, 30, 36, 57
Mitosis, 30, 31

N

Nucellar embryo, 30
  cell pattern, 37
  cell wall, 36
  development, 4
  polarization, 36, 40
  time, 33
  ultrastructure, 37
Nucellar embryony (adventive embryony),
    5, 6, 52, 63–70
Nucellus (megasporangium), 33
  crassinucellate, 34, 44
  tenuinucellate, 35
Nucellus culture, 46
  embryoid differentiation, 46
  initial cell of the embryoid, 46
Nuclei(us), 12
Nuclei(us), of embryocytes, 24

O

Ovule, 1, 16, 24
  collapse of, 22
  culture, 46
  sterile, 19

P

Parthenogenesis (ovogen apogamy,
    apozygoty, unreduced parthenogen-
    esis, reduced parthenogenesis,
    pseudogamy, etc.), 5, 11, 58, 63–70
  classification, 61
  diploid (unreduced), 9, 21, 61
  haploid (reduced), 10, 61
Plasmodesmata in cell walls, 31, 57
  of embryocytes, 26, 30

of embryoids, 45
of integumentary embryos (adventive
    embryos), 26
of nucellar embryos (adventive embryos),
    26, 36
Plastids, 25, 30, 36, 57
Polar nuclei (central cell), 22, 23, 33
Pollen grain(s) (male gametophyte)
  development, 4, 15
  fertility, 13, 15, 16, 22, 52
  germination, 15
  mature, 16
  sterility, 15, 32
Pollen tube, 15, 33
  entry in embryo sac, 15, 21
  growth, 21–22
  growth abnormalities, 21
Pollination, 3, 10–11, 32, 62
  in apomictic plants, 11
  dependence of nucellar embryony on, 5
  in plants with nucellar, integumentary
    (adventive) embryony, 32, 33
Polyembryony, 3, 8, 41–43
  adventive (nucellar, integumentary), 9
  classification, 9
  gametophytic, 9
  integumentary, 12, 41
  nucellar, 11
  sporophytic, 9
Polysomes, 30, 36
Proembryo, 37
Proembryo, zygotic (sexual), 9
Protein bodies, 36
Protoplast (cytoplasm), 30
Pseudoapospory (diplospory), 59
Pseudogamy (parthenogenesis), 62

R

Reduced generative parthenogenesis
    (parthenogenesis), 61
Reduced male parthenogenesis (androgen-
    esis), 62
Ribosomes, 30, 36, 57

S

Seed(s), 43, 51
  development, 59
  presence of adventive embryos in, 3
  reproduction, 7, 52
Seedlings, 8
  adventive (nucellar, integumentary), 8
  zygotic, 8
Semiapospory (diplospory), 6, 59

Semigamy (hemigamy), 9, 11
Somatic apospory (apospory), 5, 6
Sporophytic multiplication, 8
Starch, 24, 27
Suspensor, 37
  of adventive (nucellar, integumentary)
      embryos, 38
  of sexual embryos, 39
  variation in organization, 38

U

Unreduced male parthenogenesis (andro-
    genesis), 62

Unreduced somatic parthenogenesis
    (parthenogenesis), 61

V

Vacuoles, 25, 36
Vegetative multiplication, 6, 57
Vegetative multiplication, classifications, 5

Z

Zygote, 3, 24, 37
  cell wall, 31, 57
  polarity, 32